S0-AAL-526

FLESH MACHINE
Cyborgs, Designer Babies, and
New Eugenic Consciousness

AUTONOMEDIA NEW AUTONOMY SERIES

Jim Fleming & Peter Lamborn Wilson, Editors

FLESH
MACHINE
Cyborgs, Designer Babies, and
New Eugenic Consciousness

CRITICAL ART ENSEMBLE

AUTONOMEDIA

Acknowledgments

CAE would like to thank Nell Tenhaaf, Camilla Griggers, Simon Penny, María Fernández, Csaba Toth, John Sturgeon, G. Roy Levin, Miwon Kwon, Doug Ashford, Julie Ault, Martin Beck, Tim Druckrey, Christian Hoeller, Stelarc, Konrad Becker, Marie Ringler, Matt Fuller, Heath Bunting, Graham Harwood, Pit Schultz, Geert Lovink, Diana McCarty, V2, Merja Puustinen, Mark Dery, Reinhard Braun, Sandy Stone, Art in Ruins, Peter Lamborn Wilson, and Franco Berardi for their helpful ideas, suggestions, and comments.

CAE would especially like to thank Steven Englander and Jim Fleming for their help in coordinating this project, and for all their editorial assistance. And finally, CAE offers tremendous thanks to Faith Wilding for her editorial assistance and her illuminating drawings.

Anti-copyright 1998 Autonomedia,
Critical Art Ensemble
This book may be freely pirated and quoted.
However, please inform the authors and publisher
at the address below.

Autonomedia
POB 568 Williamsburgh Station
Brooklyn, NY 11211-0568 USA

Fax/Phone: 718-963-2603
www.autonomedia.org

Printed in the United States of America.

Contents

Introduction

The sociologist Max Weber identified the historical tendencies
and cultural trends of the 20th century as those primarily
guided by a process of rationalization. Unlike many other
social thinkers of his day across the political spectrum who
believed that increased rationalization was the process
underlying Western progress, Weber could only agree to
the extent that it would culminate in a society more
complex than any previous social order. Beyond this point,
Weber split radically from the social theorists of progress in
believing that the apocalyptic sacrifice of human well-
being inherent in the process would create a state of general
misery that would also exceed any that had come before.
The idea of reducing cultural dynamics and human activity
to a principle of instrumentality for the sake of increased
political and economic efficiency is a disturbing notion

when considered in the light of the hope for happiness and satisfaction in an age obsessed with management systems. Certainly, the hostile management systems which currently structure people's everyday lives—production management, urban management, resource management, leisure management, and expression management, to name but a few—all require bureaucratic administration. These systems (whether corporate, governmental, or military) solidify the boundaries between segments of the division of labor (whether based on race, ethnicity, gender, or class), and cause a profound sense of separation and alienation in individuals who share no solidarity beyond their common rationalized social roles and ranks in society. The feelings of alienation continue to intensify in proportion to the degree of isolation one feels from those outside one's social segment.

Weber's notion of an "iron cage of bureaucracy" that exists as a construct of purely functional imperatives (the maintenance, expansion, and perpetuation of its structure) has come to pass, so it is little wonder that the only representation of the social that seems to have any descriptive or explanatory value is the machine system. The two key machinic schemata that have captivated the minds of resistant cultural producers are the war machine and the sight machine. These intersecting liquid maps have been used to build an understanding of the structure and dynamics of the current sociopolitical management apparatuses, and in the best of circumstances have been used to develop strategies to resist the hyper-rationalized state of pancapitalism. In this work, Critical Art Ensemble (CAE) offers a contribution to the development of a third machinic map—the flesh machine. The flesh machine is a heavily funded liquid network of scientific and medical institutions with knowledge specializations in genetics, cell biology, biochemistry, human reproduction, neurology, pharmacology, etc., combined with nomadic technocracies of interior vision and surgical development. The flesh machine intersects at many points with the other two machinic systems, yet it also has an autonomous sphere of action and its own particular agenda. It has two primary mandates—to completely invade the flesh with vision and mapping technologies (initiating a program of total body control from its wholistic, exterior configuration to its microscopic constellations), and to develop the political and economic frontiers of flesh products and services.

While the war machine and the sight machine are useful devices for understanding the management of the structure and dynamics of social environments, they are less helpful for guiding resistant vectors through the new interior frontier of body invasion (the autonomous space of the flesh machine). The body is on the verge of being placed under new management, and like all exterior cultural phenomenon, it will be made to function instrumentally so that it may better fulfill the imperatives of pancapitalism (production, consumption, and order). Currently, pancapitalist power vectors' attempts to inscribe these imperatives directly onto the code of the flesh are initiating a new wave of eugenics. Under this new bio-regime, physical perfection will be defined by an individual's ability to separate he/rself from nonrational motivation and emergent desires, thus increasing he/r potential devotion to varieties of political-economic service to perpetuate the pancapitalist dynasty.

Just when it seemed that eugenics could not return to the forefront of the social arena, it appears once again, although its spectacle has been modified to suit the times. Eugenics, at least on the surface, is only implicitly attached to issues of race improvement or gene pool cleansing. Now it hides under the authority of medical progress and the decoding of nature. This latter association is primarily what makes it palatable once again. The Western intellectual history of the 20th century is marked by an obsession with coding, whether with cultural codes (linguistic or iconic representation), machine codes, or biological codes. Power vectors have been swift to take advantage of this transition to once again begin the project of totally rationalizing organic territories. The last terrestrial frontier is about to fall to pancapitalist authority.

In order to further distract the literate nonspecialist public from the development and ideological inscription of this new technological apparatus, the sight machine has consistently directed public perception toward new developments in telecommunications. Of particular importance is the idea that new telecommunication technology will make the body (if not the entire organic world) superfluous, and that human organics will "evolve" into a posthuman state of being. If such a belief is successfully deployed, questions about the goals and intentions of flesh rationalization become unimportant. After all, if the body is being done away with anyhow, and the Cartesian dream of freeing consciousness from the dead weight of the organic is on the verge of coming to pass, who cares what is done to the body or what becomes of it? Unfortunately, no virtual utopia is awaiting, nor is one even in an advanced stage of development. The current techno-

revolution is designed to keep the body, but in a redesigned
configuration that helps it adjust to the intensified rigors
of pancapitalist imperatives and to adapt to its pathological social environment. Once the goals of the flesh machine
are factored into the machine world equation, little doubt
is left that power vectors have no desire to abandon or even
undermine their material empire—all that is desired is
better, if not *total* control over their dominion.

Perhaps the greatest problem revealed in this collection of essays
is the difficulty of knowing what to do to resist the current
body invasion. While there is rapidly growing critical
knowledge about the development of the flesh machine,
strategic and tactical plans of resistance are few. Since the
flesh is a frontier zone in the development of pancapitalism,
and the situation and apparatus of invasion change with
every passing moment, strategic commitment requires a
very radical gamble on the part of resistant forces. Skepticism among specialized sectors of resistance regarding the
quality of the critique of the flesh machine is running
rampant, and to make matters worse, there is no significant
history of resistance to organic colonization to look back
upon (only scattered traces) to make possible an educated
guess about the probability of strategic or tactical success.
In addition, by attacking the flesh machine, which has
been presented as a progressive boon to humanity, the
attacker is immediately put in the position of a neo-luddite.
Science and technology in and of themselves are not the
problem, nor have they ever been. The real problem is that
science and technology are developed, deployed, and controlled by the predatory system of pancapitalism. The
mainstream development of knowledge and technology is
guided by increased efficiency in militarized production of

violence and/or by potential corporate profits in civilian markets. If a scientific producer cannot demonstrate a connection with at least one of these two possibilities, little if any investment in scientific initiatives will be forthcoming. Pancapitalism has yet to offer any sense of science in the public interest, and shows no sign of doing so in the future.

Unfortunately, we are left with only more description and more critique. At best, narratives counter to the "official story" are being created. Counter-spectacle aimed at the nonspecialist public is a significant step forward, since the mandates and methods of the flesh machine are kept as far away as possible from the scrutiny of the nonspecialist public, and because it is a step beyond the narratives of the bureaucratic ethicists whose teeth are not even the quality of dentures. However, since the flesh machine no longer works solely within the realm of the production of violence and has shifted its strategic fulcrum to the realm of seduction, much more is needed from resistant forces. What is needed is still the most elusive of all things to conjure, since this circumstance of resistance requires that the unspeakable be spoken and that the impossible be done.

1

Posthuman Development in the Age of Pancapitalism*

For the first time in history there is one globally dominant political economy, that of capitalism. Under this regime, individuals of various social groups and classes are forced to submit their bodies for reconfiguration so they can function more efficiently under the obsessively rational imperatives of pancapitalism (production, consumption, and order). One means of reconfiguration is the blending of the organic and the electromechanical. Potentially, this process could result in a new living entity distinct from its predecessors. This process, now termed posthuman development, is in its experimental stages, which in turn has lead to speculations and theories about what form this new being will take and about its probable functions. The two entities of posthuman

---

*This article was originally published in *Muae*, No. 2.

existence most commonly postulated by cyber-visionaries, techno-critics, and machine designers over the past twenty years are the cyborg and downloaded virtual consciousness. While robots, androids, and artificially intelligent machines are also generally considered part of the posthuman family, they do not emerge directly out of human organics, and hence constitute a different line of development. Cyborgs and virtual consciousness, on the other hand, are dependent upon human individuals who desire or are condemned to interface with the machine. The cyborg is a being which typically has an organic platform integrated with a complex technological superstructure; virtual consciousness is the transference of being into digitized form so that it can exist in immersive informational landscapes. This latter vision of the posthuman is one in which the Enlightenment principle of increased domination of the mind over the body moves toward full realization of complete organic erasure.

The posthuman condition is still only a potential, since fully integrated, first-order cyborgs (in which the organic platform and technological superstructure are completely interdependent) are still on the cultural horizon, and virtual consciousness is at best an entertaining speculation. Yet, both of these posthuman possibilities are already having a dramatic social impact. While virtual consciousness acts as a mythic validation of the Age of Reason, second-order cyborgs (organic infrastructures with removable, integrated technological systems) are a common actuality. This situation often leads to the conjecture that the cyborg will be the step inbetween organic life and virtual life. However, when posthuman manifestations are taken out of the context of sci-fi speculation, and placed within the specific social and economic context of pancapitalism, a much different scenario emerges. While cyborg research is moving at top velocity, research into virtual reality (VR) is moving very slowly by comparison, and the research that is being done does not aim to develop a posthuman environment nor to create a posthuman entity; rather, this work is to fortify the pancapitalist dynasty in physical space by serving both spectacular and military apparatuses. The current functions of VR, as well as the limited research into its varied potentials, are indications that virtual consciousness is not a desirable posthuman condition from the perspective of primary power vectors of the current political economy.[1]

Imaging Technology Divided

Currently, imaging technology is deeply divided between the photographic and the postphotographic—one which records and one which renders. This division corresponds very well to Althusser's division of the social into the Repressive State Apparatus (RSA) and the Ideological State Apparatus (ISA). When placed in this theoretical matrix, the actual functions of immersive technology become intelligible, and indicate that inserting disembodied consciousness into virtual systems is not a current or future strategy for the development of posthuman entities by investing agencies.[2]

The ISA is the structure of control maintained through the manipulation of a given culture's symbolic order. Through this structure, ideology is deployed, reinforced, and morphed in order to maintain the integrity of the given power structure. When enveloped in the ISA, an individual's perceptions, thought structures, and behaviors are molded to varying degrees so as not to conflict with dominant historical and socioeconomic imperatives and conventions. The family, schools, the church, and the media function as the pedagogical institutions through which ideology replicates itself. While family and religion were dominant in precapitalist society, in pancapitalist society, the education system and the media have become the dominant institutions of socialization, causing family and church to fall to a secondary support position. Of particular interest here is that communication technologies have been of tremendous aid in empowering these latter institutions and in causing this pedagogical shift to occur. The ideologically charged and immersive representational environment generated through the use of mass imaging technologies is called spectacle.

The infrastructural counterpart of the ISA is the Repressive State Apparatus (RSA). It is the total structure of control maintained through violent, militarized social intervention. Its functions are to defend the social system from invasion, to protect "investments" in alien social systems, and to maintain internal social order. Institutions participating in this apparatus include armed services, national guards, police, prison systems, and intelligence and security agencies. Like the dominant institutions of the ISA, the RSA depends upon complex technology in order to remain effective; however, unlike the ISA, where imaging technologies are an end in themselves, the RSA requires a full array of weapons technology integrated with vision technology in order to carry out its mandate of repression through violence.

The imaging technology of the RSA is generally superior to that of the ISA, but the functions of the technology for each system are so vastly different that concrete comparisons are very difficult. The unique characteristics of the technology for each system are many and make for highly contrasted systems. The first significant issue of difference is visibility. The imaging engines of the ISA are relatively open to public scrutiny. In fact, for these imaging systems to work, the technology which delivers its message (radios, televisions, cinema, etc.) must be widely deployed and/or made accessible. While the machinery that actually generates the spectacular images is typically kept out of the representational frame, that machinery too can and does become one of the objects of spectacle. Passive media participants are familiar with the existence and function of many components of media systems, including cameras, control rooms, editing suites, communications satellites, film studios, and so on. This is true to the extent that imaging systems are often the object of the jubilant celebrations of technology. For example, who can deny the public excitement generated over a cable system delivering 500 channels, or the public enthusiasm for surfing the World Wide Web? Couple this situation with media stunts engineered to further ingratiate new technology to the public, and the celebration becomes even more intense.

For example, NASA's use of the robotic walker Dante to record events in a volcano was basically a useless endeavor in terms of scientific data collection. NASA has very little use for volcanic data. However, this project did serve two functions: First, it acted as a public relations campaign to demonstrate in very dramatic terms the current state of hi-tech development. (In actuality, Dante was produced with robotic technology that was developed many years ago, but the data acquired sure seemed impressive, as did its descent into the volcano. The snapping of its retrieval cord at the end of the mission was a little embarrassing, but it was not highlighted). The second and primary reason for the Dante mission was to develop software for autonomous robots, which could then be used for other purposes. In addition to building a greater public appreciation for the technology itself, stunts such as this one also help to construct a heroic image for those who use complex technology. Certainly, this is part of the reason for the current series of multinational space missions. In all, this kind of celebratory spectacularization is great news for cyborg development, as people increasingly desire to be close to complex technology. One certainly cannot help but wonder if new technology is not the most important object of spectacle at this moment.

For the RSA, technology is generally kept hidden, and at least during peacetime, it is no cause for celebration. This is not to say that military technology is not spectacularized, for it most certainly is. The public display of military technology is done primarily for purposes of disinformation. When a new weapon or imaging system is developed, sometimes it is wise to let others know that it exists; however, what must be kept secret are the precise parameters of what it can do. Such incomplete information leakage constructs an inflated sense of security in friendly populations, and a sense of fear in enemy populations. It also entices enemy powers to overestimate the capabilities of the system. For example, during the early years of the cold war, US military officials grossly overestimated the numbers of Soviet nuclear missiles; this caused not only a panic in the military over an assumed "missile gap," but also caused a panic in the civilian population. This situation was not corrected until data were retrieved from early spy satellite reconnaissance missions in the early 60s which showed that far fewer missiles existed than US intelligence orginally believed. A more contemporary example is the spectacularization of the Patriot Missile System during the Gulf War. As it was presented in the media, this piece of techno-junk required divine intervention for it to function. Even after the system was a proven battlefield failure, it was still presented as the ultimate anti-missile defense system, simply by either constantly replaying the footage of its few successes, or by showing images of it not working accompanied by an authoritative voice-over saying that it was working.

The compelling point here is that spectacular engines employed for this duty are functional, because they use a postphotographic model in which imaging systems are totalizing rendering devices. The images produced under this model, and those presented within the media context, are inherently untrustworthy, and flow within the representational fictions of realistic illusionism. While the engines and methods of production do not call attention to themselves within the screen's frame, it is common knowledge that media images—from fantasy cinema to nonfictional newscasts—are engineered and designed by a plethora of means too numerous to list in this essay. This skeptical view has become so common that conspiracy theorists can even claim that the moon landing was a US government hoax, and can do so with a modest amount of credibility. Paradoxically, and in a sublime moment of doublethink, the public implicitly understands the rendering hoax of postphotography, yet still often finds spectacular images credible.

Contrary to the spectacular model of rendering is the more traditional model of photographic imaging systems, which focuses on principles of recording. While spectacular images created by the ISA for the public continually slide, dissolve, and recombine, images produced by the RSA *for its own use* are relatively stable, if the imaging systems function as intended. RSA images are produced for the purposes of mapping and/or surveying territories and populations: They have a material referent, which is validated through practical application based on information extracted from the image. The goal is to produce images in which the map is to the greatest extent possible a representational equivalent of the territory. The higher the accuracy of the representation in relation to the designated referent, the higher the value of the image. This strategy is based on the understanding that that which can be visualized and mapped can be controlled, as long as this vision system is integrated with an array of effective weapons systems (both human and electromechanical) that can be deployed in a contested territory or among a resistant population.

The need for accurate images is not just a matter of strategic necessity, but also a matter of cost-effective control. The RSA is an incredible drain on financial resources when left standing, but it costs even more to use its forces (although it must be noted that wartime economy can be profitable). Given that the grand majority of contemporary first-world military conflicts fall into the category of "police action" (which is also the least profitable of military activities, as police actions maintain the empire rather than expand it), military deployment and use must be as precise as possible. The key distinction between images produced by the RSA and those produced by the ISA that arises out of this instrumental imperative is that RSA images serve a pragmatic military function, while those of the ISA serve an aestheticized sociopolitical function.

To return to the issue of secrecy in regard to RSA activities, the pragmatic structure of its vision can maintain its integrity only as long as the specifics of the vision engines and the images they produce are kept classified. Once the images are recontextualized in the image barrage of the ISA, they fall victim to aestheticization, lose their usefulness, and become obsolete. This obsolescence typically occurs when one generation of vision technology is replaced by a new generation of more accurate technology, and creates the opportunity for technological spin-offs for the realm of spectacle. Computers, the Internet, and communication satellites are all examples of representational engines that lost some or all of their value as RSA systems, and hence no longer had to be kept monopolized. These engines were reconfigured and redeployed for the purpose of producing spectacle, which in turn indicates that there is a vast intersection between processes of aestheticization and obsolescence, which in turn further suggests that the much-celebrated postphotographic principle of rendering is often still in the service of its parent, the photographic model of recording.

This is not to say that all institutions for image production fit neatly into either the RSA or the ISA. The centralization of capital in various multinational industries allows the development of vision engines with double functions serving both the ISA and the RSA. These engines are often developed independently by institutions that intersect both apparatuses. For example, the institution of medicine plays a pivotal role in both the ISA and the RSA. While on the one hand it participates in the ISA by producing a spectacle that dictates what is physiologically and psychologically "healthy" and "normal," it also participates in the function of the RSA through its ability to muster forces to support these standards, and through its administration of a system of institutions in which "deviants" may be imprisoned (asylums, hospitals, rehab clinics, halfway houses, etc). At the same time, this industry is still partly dependent on the RSA, since some of its technology comes from obsolete machines released to friendly institutions. For example, telepresent robotic surgery is now being developed by the military for use in the field. One can be fairly certain that before this med-tech begins to trickle down to civilian medicine, it will have become a fully functional military option, and that by that time the military will have moved onto newer and better options for field surgery.

As the border between the ISA and RSA grows increasingly cluttered with relative independents, the technological state-of-the-art starts to drift back and forth between spectacular processes of aesthetic production and militarized processes of pragmatic production. Cutting-edge image production still favors the RSA, but the situation is becoming increasingly hazy because many imaging systems are assuming dual functions. What is certain is that rendering is the foundation for spectacular image production, and that recording is still the foundation for militarized image production.

The Dual Function of Immersive Technology

Given the theoretical matrix explained in the last section, the
likelihood of realizing the dream of VR as a liberating
future habitat for humanity seems quite remote. In fact,
VR seems to be used for every imaging purpose except as a
liberating habitat. Its use by the spectacle is minimal, as no
investing agency seems able to conceive of a useful (instrumental) application for it. Currently, VR takes a very
secondary position to older nonimmersive screen-based
systems. While the World Wide Web, the Internet, and
cable television seem to be exploding with new possibilities (both compelling and loathsome), VR is beginning to
stagnate. Its position is limited to arcade entertainment
and to secondary-display technology that helps boost
consumption. One example of this latter variety of application is the use of VR in some department and furniture
stores in Japan. A shopper can enter a virtual environment
and (within the limits of the product line) render a desired
domestic environment to see if it meets with he/r expectations before purchasing the needed merchandise. If he/r
virtual vision does not meet he/r expectations, s/he can
redesign the space until it does. The buyer is thus given
extra assurance that s/he will get what s/he wants. Obviously, a system like this functions only when there is a
variety of purchasing options, when the object of consumption cannot be physically displayed, and when the
purchase is costly. Hence this application has very limited
spectacular use. Further, this application is only one small
step beyond the use of X-ray machines in shoe stores back
in the 30s and 40s. The shopper could X-ray he/r foot to
make sure the shoes about to be purchased were a perfect
fit. In terms of the spectacle of consumption, the real

problem for VR is that there are very few occasions when the institutions selling the products want to give even the smallest amount of authentic choice to the consumer.

The infinite choice and total control promised by VR are precisely the options that investment institutions want to avoid, and hence, they are not going to pursue VR technology with any vigor until someone is able to negate its liberating logic. This is also why investment capital is flowing overwhelmingly in the direction of screen technology, such as the World Wide Web. (The rocketing prices of shares of companies like Netscape and Yahoo when they went public clearly indicate the flow of capital). On the Web, the producer of the page controls the rendering process. While this element of Web production seems to favor the cyber-individual, and accounts for much of the celebration of the Web, institutions are aware that those with the greatest amount of capital can use the latest software and state-of-the-art trained labor to achieve maximum novelty and aesthetic seduction, overwhelm competitors for visibility through additional advertising in a variety of media, and offer additional incentives (usually chances at prizes or free merchandise) for using the page. (And, if consumers are willing to give personal information for market research to increase their chances of getting these incentives, so much the better). If the lure is carefully constructed, the advertiser can expect to monopolize a Web consumer's time. Interactivity in this case means the ability of the consumer to view a product, purchase it, and/or move onto other purchasing opportunities in the given product line. This is the kind of spectacular technology that pancapitalist ISA will support, not just with investment, but also with legislative and regulatory support.³ Technologies which truly offer emergent choice and devalue centralized economic control are not worth an investment. Currently, the posthuman has no place in VR, and VR has a very small material place in the ISA.

VR's primary value to the ISA is not as a technology at all, but as a myth. VR functions as a technology that is out on the horizon, promising that one day members of the public will be empowered by rendering capabilities which will allow them to create multisensual experiences to satisfy their own particular desires. The mysterious aura constructed around this technology associates it with the exotic, the erotic, and potentially, with the ethereal. By perpetuating the myth of a wish machine that is always about to arrive, the pancapitalist ISA builds in the population a desire to be close to image technology, to own it. Unfortunately, most technology is being designed for precisely the opposite purpose from that of a wish machine, that is, to make possible better control of the material world and its populations. This combination of myth and hardware sets the foundation for the material posthuman world of the cyborg.

The RSA proceeds along a different route. All the potentials of VR are being used to create more accurate simulators. However, the core of this immersive image is based on recording. Usually, the technological environment which the VR system is designed to simulate has already been built or at the very least is under construction. As to be expected, the virtual image again has a very clear material referent. For example, a fighter jet simulator attempts to replicate the interior technological environment as accurately as possible. The quality of the replication is judged practically by how well a pilot trained in the simulator does in the actual cockpit. The exterior virtual environment in which the simulated technology functions makes use of both recording and rendering. However, recording is still dominant, as the trainers attempt to place trainees in specific rather than in general environments. Returning to the example of the jet fighter simulator, the pilot is placed in an environment closely resembling the one in which s/he will be flying. The ground, anti-aircraft batteries, and enemy planes are rendered as accurately and as specifically as possible based on recorded photographic images and intelligence data, whereas more random variables, such as atmospheric conditions, will be rendered in accordance with generalized configurations.

As with the imaging systems of the ISA, the goal is not to prepare a person for life in the virtual, but to specify, regulate, and habituate he/r role in the material world. Virtuality has no independent primary function in the RSA; rather, it has a dependent secondary support function. What is really odd about this situation is that the mythic gift of VR—complete control of the image—is negated. The virtual images are completely overdetermined by specific configurations in the material world. The limited evidence available to the public indicates that no preparations are being made for immersive virtual information warfare. This possibility seems limited to the screenal economy of cyberspace. However, since RSA activities are classified, plenty of room exists for conspiracy theorists to speculate. At the same time, given current trends in investment, research, and development, combined with the very clear imperatives of pancapitalism, such speculations have only a very modest amount of credibility.

Preparing for Posthumanity

If the habitat of VR and the virtual entity are eliminated as practical categories of the posthuman, the only possibility left is the cyborg. In terms of social perception in technologically saturated economic systems, being a first-order cyborg covers a broad range of possibilities, ranging from a desirable empowering condition to an undesirable, dehumanizing one. However, there is plenty of time for spectacle to sort out differing perceptions of the first-order cyborg. Cyborg development is moving at a pace which allows adequate time for adjustment to the techno-human synthesis. Currently, the process is in very different stages at specific institutions. For example, the military has advanced furthest, and has developed a fully integrated second-order cyborg, while corporate and bureaucratic institutions are meeting with reasonable success in their attempts to convince workers of the need to meld body and technology.

Within many civilian social institutions, cyborg development is progressing cautiously enough that members have a difficult time knowing what a cyborg is, perceiving one, or realizing that they are being transformed into one. Is a cyborg any person who has a technological body part? Does having an artificial limb or even contact lenses place one in the category of cyborg? In a sense, the answer is yes, as these pieces of technology are integrated with the body, and the individual is relatively dependent upon them. However, in terms of posthuman discourse, the answer is probably no, as there is only a simple engineered interface between the technological and the organic. The posthuman model that seems to be developing is McLuhanesque—that is, the techno-organic interface should extend the body beyond the fluctuating degree zero of everyday normalization. What is spoken about in the case of artificial limbs or contact lenses is the means to make the body, to the greatest extent possible, conform to "accepted" social standards. What is interesting about precyborgian technological additions to the body is that one key ideological imperative having a direct affect on posthuman development begins to show itself—body-tech is valued as means to better integrate oneself into the social.

Another common question is whether radical technological body intervention, such as gender reassignment, makes one a cyborg. Obviously, since such procedures are only organic recombinations devoid of technology, they fail to create a cyborg class being. However, these interventions do play a role in cyborg development, because they continue to prepare specific publics to perceive these operations as normal and often desirable. This is particularly true of interventions done solely for aesthetic purposes. The social "abnormality" of organic decay acts as an ideological sign that channels people toward the consumption of services for body reconfiguration, to enable them to best fulfill the social imperatives of body presentation in pancapitalist society. What is truly important about this development is that technological intervention disconnected from issues of deviance, sickness, and death is being normalized. Extreme body invasion as a socially accepted practice is a key step in cyborg development.

Military and Civilian Cyborgs

There is no need to dwell on the development of a second-order military cyborg. The only surprise here is that it took so long to happen. From the common grunt to the heroic jet fighter pilot, the military conversion of humans to cyborgs has become a necessity. The Hughes Corporation has successfully developed a custom-fitted techno-organic interface for the infantry which offers an integrated system of vision, communication, and firepower. Soldiers are no longer soldiers; as the military says, now they are "weapons systems." The posthuman has announced itself in a happy moment of military efficiency. However, the infantry "weapons system," while actual and functioning, is a minor interface when compared to the developing "Pilot's Associate" (McDonnell-Douglas). In addition to having a state-of-the-art interdependent pilot/machine interface (unless the machine thinks that the organics are failing, and it must take over the mission), the "Pilot's Associate" offers Artificial Intelligence (AI) support analysis in mission planning, tactics, system status, and situation assessment. Here we find a clear indication of what body "enhancement" is going to mean in the age of the posthuman. Body enhancement will be specific to goal-oriented tasks. These tasks will be dictated by the pancapitalist division of labor, and technology for body modification will only allow for the more efficient service of a particular institution.

Unfortunately for the multinationals, the development of the civilian cyborg has not moved along as quickly. Since the civilian sector does not have the advantage of telling its forces that being-as-cyborg will prolong one's life in the field, corporate power vectors are still deploying ideological campaigns to convince civilians of the bureaucratic and technocratic classes that they should desire to be cyborgs. The spectacle of the civilian cyborg moves in two opposing directions. The first is the utopian spectacle. The usual promises of convenience, access to knowledge and free speech, entertainment, and communication are being trotted out by the usual media systems with varying degrees of success; but anyone who has paid attention to strategies of manufacturing desire for new technologies can read right through the surface of these codes. Convenience is supposed to mean that work becomes easier, and is accomplished faster; in turn, this means that individuals work less and have more free time because they work more efficiently. What this code actually means is that the workload can be intensified because the worker is producing more efficiently. Entertainment and information access are codes of seduction that really mean that individuals will have greater access to consumer markets of manufactured desire. Better communication is supposed to mean greater access to those with whom an individual wants to communicate. The actuality is that agencies of production and consumption have greater surveillance power over the individual.*

---

*The deployment of utopian promise is commonly used to form future markets. For additional information on the development of utopian promises associated with telecommunication technologies, please see the Appendix.

In contrast to utopian spectacle is the spectacle of anxiety. The gist of this campaign is to threaten individuals with the claim that if a person falls behind in the technological revolution, s/he will be trampled under the feet of those who use the advantages of technology. This campaign recalls the socioeconomic bloodbath of the ideology of Social Darwinism. The consumer must either adapt or die. From the perspective of pancapitalism, this campaign is quite brilliant, because unlike the military (where the soldier is supplied with technology to transform he/rself into a weapon system), the civilian force will buy the technology of their enslavement, thereby underwriting a healthy portion of the cost of cyborg development as well as the cost of its spectacularization.

The current spectacle of technology is having an effect on the civilian population of the appropriate classes, although cyborg development in this sector is a little more subtle than in the military. Most people have seen the first phases of the civilian cyborg, which is typically an information cyborg. They are usually equipped with lap-top computers and cellular phones. Everywhere they go, their technology goes with them. They are always prepared to work, and even in their leisure hours they can be activated for duty. Basically, these beings are intelligent, autonomous workstations that are on call 24 hours a day, 365 days a year, and at the same time can be transformed into electronic consumers, whenever necessary.

In this phase of posthuman development, the will to purity, explicit in the spectacle of anxiety, manifests itself in two significant forms: First is the purification of the pancapitalist cycle of waking everyday life. Cyborgs are reduced to acting out rational, pragmatic, instrumental behaviors, and in so doing, the cycles of production (work) and consumption (leisure) are purified of those elements deemed nonrational and useless (by the pancapitalist system). It seems reasonable to expect that attempts will be made to reduce or eliminate regenerative, nonproductive processes like sleeping through the use of both technological and biological enhancement. The second is a manifestation of ideological purity which persuades the cyborg to obsessively value that which perpetuates and maintains the system, and to act accordingly. The prime disrupter of this manifestation of purity is the body itself with its endlessly disruptive physical functions, and the libidinal motivations inherent in human psychology. Hence technological advancement alone will not create the best posthuman; it must be supported by developments in rationalized body design.

Final Preparations for Posthumanity

The military has long understood that the body must be trained to meet the demands of its technology. Consequently, it puts its organic units through very rigorous mental and physical training, but in the end, it is clear that this training is not enough. Training can only take a body to the limits of its predisposition. Pancapitalism has realized that the body must be designed for specific, goal-oriented tasks that better complement its interface with technology within the real space of production. Human characteristics must also be rationally designed and engineered in order to eliminate body functions and psychological characteristics that refuse ideological inscription.

The mature appearance of the flesh machine is perhaps the greatest indication that the magical data dump of consciousness into VR is not being seriously considered. If it were, why invest so heavily in body products and services? Conversely, why should capital refuse an opportunity that appears to be the greatest market bonanza since colonization? Digital flesh is significant in the mapping of the body, but its value depends upon the practical applications that are derived from it; these, in turn, can be looped back into the material world. The body is here to stay. Unfortunately, the body of the future will not be the liquid, free-forming body which yields to individual desire; rather, it will be a solid entity whose behaviors are fortified by task-oriented technological armor interfacing with ideologically engineered flesh. Little evidence is available to indicate that liquescence is different in postmodernity from what it was in modernity—the privilege of capital-saturated power vectors.

Notes

[1] Vague terminology, such as CAE's use of the term "power vectors," has unfortunately become a necessary evil in the description of pancapitalist political economy. The dynamics of domination are at present impossible to concretely identify and describe because the flow of power moves and shifts at such an extreme velocity that it cannot be located at a fixed point where it can be empirically studied. Please see *The Electronic Disturbance*, Chapter 2 for an extended discussion of this problem.

[2] Here, it must be noted that CAE is not attempting to make an apology for either/or structuralism, in which all social phenomena fit neatly into a binary package; rather, CAE is using the division as a liquid continuum along which countless inbetween, hybrid, and recombinant possibilities occur, in a constant state of transformation over time. CAE also believes that the boundaries of a given ISA/RSA are extremely fuzzy. In the age of pancapitalism, no national or cultural borders are rigid enough to allow the establishment of a concrete unit of analysis. We are therefore limited to discussing only first-world political and economic trends and tendencies.

[3] Since political agencies do not want to disrupt the WWW marketing bonanza by offending consumers, regulation is proceeding at a slow and cautious pace. For example, the steps being taken to limit Web content (such as legislation to control "obscenity" or to eliminate information on weapons construction) are presented to the public as security measures. These opening attempts to regulate the Web also function as preliminary research to discover or invent the best means to enforce regulation. Like media of the past, such as radio or cinema, totalizing regulation will not appear until the fiscal structure of cyberspace is firmly in place. In addition, regulating agencies must wait until use of the Web and the Internet become a necessary part of everyday life for individuals of higher class rankings, in order to minimize resistance to regulatory acts.

2

Nihilism in the Flesh

While much of the current cultural discussion regarding technoculture focuses on issues emerging from new communications technology, there is an exponentially growing interest in and discussion of flesh technology. Like the discussion of new communications technologies, this discourse vacillates wildly from the intensely critical and skeptical to the accepting and utopian. However, the most significant intersection between the two discourses is their parallel critique of vision enhancement. Whether it is the development of global satellite vision or the development of micro interior vision, imaging systems are key to both apocalyptic or utopian tendencies. For example, sonography can be used to map an ocean floor, or it can be used to map uterine space. In both cases, such an imaging system functions as a first step toward the ability to culturally engineer and ideologically design those spaces. As these two spheres of technology continue to intermingle, a recombinant theory of the relationship of populations and bodies to technology has begun to emerge that conflates theories of the social and the natural. The existence of such theories under the legitimizing mantle of the authority of science is not new, and in fact the theories have fallen in and out of favor since the 19th century. They continually re-emerge in different guises, such as Social Darwinism (Malthusian and Spencerian philosophy), eugenics, and sociobiology. In each case the results of such thinking have been socially cata-strophic, setting loose the unrestrained deployment of authoritarian ideology and nihilistic social policy.

Apparently theories of *deep* social evolution have come into favor again, and are rising from the grave to haunt unsuspecting populations. Socially dangerous principles of cultural development, such as fitness, natural selec-tion, and adaptability, are again in fashion. Consider the following quote from the announcement for the 1996 Ars Electronica Symposium and Exhibition (Ars Electronica is a very prestigious annual conference for multimedia artists, media critics, scientists, hackers, and technicians):

> Human evolution, characterized by our ability to process information, is fundamentally en-twined with technological development. Complex tools and technologies are an inte-gral part of our evolutionary "fitness." Genes that are not able to cope with this reality will not survive the next millennium.

This quote contains some of the most frightening authoritarian language since the Final Solution, and presents the threat of "adapt or die" as a value-free social given. To what is the reader expected to adapt? To the technology developed under the regime of pancapitalism for the purpose of better implementing its imperatives of production, consumption, and control. There is nothing evolutionary (in the biological sense) about the pancapitalist situation. It was engineered and designed by rational agencies. "Fitness" is a designated status that is relative to the ideological environment, not the natural environment. History repeats itself, as those resistant to authoritarian order must once again separate the cultural and the natural, and expose the horrific nihilistic tendency that arises when the two are confused.

Nihilism

Nihilism can have either positive or negative political associations. For example, some liberationists view nihilism as a revolutionary strategy capable of dissolving boundaries which retard the full exploration of human experience, while those interested in maintaining the status quo view it as a method of social disruption which manifests itself in destruction and chaos. Certainly the original description of nihilism, in Turgenev's novel *Fathers and Sons*, presented it as a revolutionary method designed to promote Enlightenment political principles. The engine of nihilism in this case was reason, and its application manifested itself in an overly deterministic and domineering model of Western science. Turgenev contrasts the nihilist position with Christian models of faith and a monarchist social order. While many who situate themselves on the left can sympathize with the nihilist's will to free he/rself from the constraints of the traditional model of church and state, there is also an uneasy feeling about this variety of nihilism, as a danger exists of replacing one tyrant with another. One cannot help but question if replacing faith and understanding with reason and knowledge could lead to an equivalent state of oppression. Nietzsche makes this point very elegantly in his assertions that movement toward purity and uncritical acceptance (in this case, of reason) always leads to hegemony and domination.

The case of Nietzsche in regard to nihilism is peculiar. While the Nietzschean notion of philosophy with a hammer seems to fit well with the nihilistic process, Nietzsche actually inverts the argument. From his perspective, the ability of humans to challenge dominant institutions is an affirming quality. It affirms life and the world. While the process has elements of conflict and destruction, acts of skepticism, disavowal, and resistance are intentionally directed toward the possibility of freedom, and thereby redeem people from the horrid fate of willing nothingness, rather than not willing at all. From this perspective, the primary example of the pathologically nihilistic will made manifest is the institution of the church in particular and religion in general. Religions encourage the subject to bring about he/r own disappearance, and thereby, to eliminate the world which envelops he/r. One abhors presence, and seeks absence. The problem for Nietzsche is that he cannot accept the principles of absence (the soul, God, the heavenly kingdom) that are dictated to society under the authority of church rule, and perpetuated by an unquestioning faith. Nietzsche demands that life rest in experience and in presence. To negate the given is an unacceptable nihilistic position that undermines humanity itself.

On the other hand, if theological principles are accepted, one can easily see how the positions of secularists appear nihilistic. To sacrifice one's soul to the immediacy of experience is eternally destructive. The immediacy of the sensual world should be understood as a site of temptation that negates the joy of eternity. Those who focus their daily activities on the sensual world are doomed to the torture of privation in this life, and to damnation in the next life. To choose an object other than God is to be continuously left unfulfilled, and during this time the soul decays from neglect. In terms of Eastern theology, the situation of subject-object is mediated by the hell of desire, which can only be pacified when the subject is erased, and thereby returned to the unitary void. In both the Western and the Eastern varieties of religious life, the subject can only find peace by affirming God (as opposed to affirming the world).

The truly interesting and relevant point here in regard to evolutionary social theory is that the 19th century conflict over the nature of nihilism has a common thread. No matter what side of the debate one favors, the discourse centers around institutional criticism. Nietzsche attacks the church and its doctrines, while the church attacks secular institutions such as science. People are not the object of nihilism, no matter how it is defined. However, when nihilism is combined with notions of social evolution, the object of nihilism (whether valued as good or bad) is people! It speaks of the fitness of some, and the elimination of others. It is not a racial construction that the authoritarians of social evolution seek to eliminate, but people of a race; it is not a class that they seek to eliminate, but people of a class; it is not an anachronistic skill that they seek to eliminate, but people who have this skill.

Evolution is a Theory, Not a Fact

To be sure, evolutionary theory has become such a key principle in organizing biological information that some toxic spillage into other disciplines is almost inevitable. It commands such great authority that its spectacle is often confused for fact. At present, evolutionary theory is primarily speculative; no valid and reliable empirical method has been developed to overcome the temporal darkness that this conjecture is supposed to illuminate. Consequently, evolutionary theory circles around in its own self-fulfilling principles. It is in an epistemological crisis, in spite of authoritative claims to the contrary.

The tautological reasoning of evolutionary theory proceeds as follows: Those species with the greatest ability to adapt to a changing environment are naturally selected for survival. Those that are selected not only survive, but often expand their genetic and environmental domains. So how is it known that a species has a capacity for adaptation? Because it was naturally selected. How is it known that it was selected? Because it survived. Why did it survive? Because it was able to adapt to its environment. In spite of this logical flaw of rotating first principles, evolutionary theory brings a narrative to the discipline that makes biological dynamics intelligible. While the theory can in no way approach the realm of certainty, it does have tremendous common-sense value. If for no other reason, evolutionary theory is dominant because no one has been able to produce a secular counternarrative that has such organizational possibilities.

Evolution is an intriguing notion for other reasons too. The idea that natural selection is a blind process is certainly a turning point in Western thinking. There is no teleology, not even the guiding "invisible hand." Instead, evolution gropes through time, producing both successful and unsuccessful species. Its varied manifestations display no order, only accident. This notion is an incredible challenge to the Western desire for rational order. At best, God *is* playing dice with the universe. The very anarchistic strength of this notion is also its scientific downfall. How can the accidental be measured in causal terms? For example, the engine of physical adaptability is mutation. If mutation is the accidental, uncommon, unexpected, and anomalous, how can it be quantified, when the knowledge systems of science are based on the value of expectation and typicality?

Can we say with any degree of assurance that social development is analogous to this model of biological development? It seems extremely unlikely that culture and nature proceed in a similar fashion. Cultural dynamics appear to be neither blind nor accidental. While the occurrence of chaotic moments in social development cannot be denied, unlike with biological evolution, they do not render the same totalizing picture. Cultural evolution, if there is such a thing, seems for the most part to be orderly and intentional. It is structured by the distribution of power, which can be deployed in either a negating or affirming manner.

Culture and Causality

The ever-changing and transforming manifestations of power over time are the foundation of what may be considered history. Power manifests itself in countless forms, both as material artifacts and ideational representation, including architecture, art, language, laws, norms, population networks, and so on, which is to say as culture itself. When considering either culture or history, it seems reasonable to contend that evolution (in its biological sense) plays little if any role in the configuration of social structure or dynamics. For example, the history of industrial capitalism spans only a brief 200 years. In the evolutionary timetable, this span of time scarcely registers. The biological systems of humans have not significantly changed during this period, nor for the last 10,000 years, and hence it would be foolish to think that evolution played any kind of causal role in the development of capitalism. In fact, humankind's seeming evolutionary specialization (a mammal that specializes in intelligence) places it in a post-evolutionary position. With the ability for advanced communication using language capable of forming abstract ideas, in conjunction with the ability to affect and even control elements of the body and the environment, humans have at least temporarily inverted significant portions of the evolutionary dynamic. In an astounding number of cases, the body and the environment do not control the destiny of "humanity";

rather, "humanity" controls the destiny of the body and its environment. Unlike the evolutionary process, social development is overwhelmingly a rationalized and engineered process.

If the proposition that social development is a rationalized process (perhaps even hyper-rationalized, under the pancapitalist regime) is accepted, can evolutionary principles such as natural selection or fitness have any explanatory value? This possibility seems very unlikely. For instance, there is nothing "natural" about natural selection. At the macro level, the populations that have the greatest probability of coming to an untimely end are not selected for elimination by a blind natural process; rather, they are designated as expendable populations. In the US, for example, the problem of homelessness exists not because there is insufficient food and shelter for every citizen, nor because this social aggregate is unfit, but because various power sources have chosen to let the homeless continue in their present state. The selection process in this case has agency; it is not a blind and accidental process. What is being selected for in the age of pancapitalism (and for most of human history) are cultural characteristics that will perpetuate the system, and maintain the current power structure. This process is intentional, self-reflexive, and at its worst, systematic—in other words, intensely rational.

The concept of fitness follows the same unfortunate trajectory. Once this concept is taken out of its original biological context and placed into a social context, its explanatory power evaporates. When the concept of fitness intersects an intentional environment, the idea is transformed from a relatively neutral one to one that is intensely value-laden. Unlike the biological concept of fitness, a category measured by the emergent manifestations of survival, the sociological concept of fitness functions as a reflection of a particular population that is then projected and inscribed onto the general population. The valued characteristics (beauty, intelligence, "normal" body configuration, etc.) that constitute fitness are designed and deployed in a top-down manner by power vectors which control social policy construction and image management and distribution. In a social environment which has solved the challenge of production, fitness has no real meaning other than to mark acceptable subjects, which in turn marginalizes and/or eliminates "deviant" subjects. Without question, when fitness is placed into a sociological context, it becomes a hideous ideological marker representing the imperatives of the political-economy that deployed it.

Nature as Ideology

Three decades ago Roland Barthes sent an illuminating flare into the political air to warn us of the socially catastrophic results of using nature as a code to legitimate social value. Under authoritarian rule, the social realm is divided into the natural and the unnatural (the perverse). Everything of value and of benefit to the empowered vectors of a given social system is coded as natural, while everything which negates its demands by prompting alternative or resistant forms of social activity and organization is coded as unnatural in the environment of representation. But this binary system is more complex. Given that one of these values of empowered vectors is that of militarization in all its forms, a nasty wound opens as the social fabric is ripped by contradictory ideological forces. On one hand, nature is viewed in a very gentle sense as moral and pure, and thereby good. Hence that which is natural is also good. On the other hand, when perceived through the evolutionary ideological filter as a realm in which only the strong survive the bloodbath of life, nature becomes abject, dangerous, and amoral. Hence, that which is natural (sovereign) must be repelled. The ideological role of the code of nature is doubled, and simultaneously exists as value and as detriment, thereby allowing the code to float from one meaning to its opposite. All that authoritarian power must do is contextualize the code, and it will speak in whatever manner is desired by the social vectors with the power to deploy it. In addition, for this code of control to function, its inherent contradiction must be flawlessly sutured. This is done through spectacular narrowcasts into the fragmented condition of everyday life.

It seems rather obvious that importing legitimized theories of natural dynamics (in the case of pancapitalism, evolutionary principles) into the ideological fabric is a necessity if this overall coding system is to function. In this manner the constructive qualities of a given regime can be coded as natural, as can its pathologically nihilistic and destructive tendencies, even, and perhaps especially, when they are aimed at other people! Thus the code truly is totalizing. It does not have to be split into a binary which has a boundary that authoritarian order cannot cross. Authoritarian power can occupy all social space with impunity, both normal and deviant, for constructive or destructive purposes.

Biohazards

When the dark code of nature (survival of the fittest) is efficiently deployed within a given population, genocidal nihilism becomes an acceptable course of social action. While the code legitimizes and masks military aggression for the purpose of acquiring territory and resources, the will to purity has been known to function as an independent parallel goal, as in the case of Nazi Germany. Currently, there is a shift in temperament; genocide is increasingly becoming less a matter of territory and resources, and more a matter of the will to purity. In the days of early capital, when the riddle of production was still unsolved, land/resource appropriations were the primary reason for genocide. The examples are, of course, well known: the kulak genocide under Stalin, or the aboriginal genocides in the US and Australia. In these cases, the will to purity (ideological in the case of the former and racial in the case of the latter), was secondary, and functioned primarily as the rhetoric and the justification for the actions. Certainly, one can expect to see more genocides typical of early capital in the third world, where for reasons of imperial design, production cannot meet the demands of the population. The same may be said for industrial nations in the process of restabilizing, as in Bosnia. However, in the time of first-world late capital with its consumer culture, global media, global markets, and product excess, direct military actions seem less necessary, because geographic territory is in the process of being devalued.

With economic expansion via territorial occupation in the process of disappearing, the will to purity (fitness) stands on its own as a prime reason for genocide. Currently, genocidal nihilism tends toward elimination of "deviant" subjectivity. This new form of nihilism is much more subtle. The day of the death camp designed for maximum efficiency is over, and in its place are prisons, ghettos, and spaces of economic neglect. By making it seem that the condition of extreme privation is a part of the natural order, rational authority can eliminate populations without direct militarized action. In some cases, the designated excess population will participate in its own destruction as individuals are forced by artificially produced physical need and environmental pressures to do whatever is necessary to acquire withheld resources. In turn, these actions are replayed by the media as representations of the dangerous natural qualities of given races, ethnicities, or classes that must be controlled. Ironically, activities and environments which were intentionally designed become representations of nature, and proof of fitness theory.

Accidental opportunities also have great potential for exploitation. In the early years of the AIDS crisis in the US, when the virus seemed to affect only gay men, IV drug users, and Haitians, the Reagan Administration exploited this opportunity to eliminate some "degenerate" populations; after all, they were unnatural, impure, and unfit. By refusing to intervene or even acknowledge the existence of the virus, the Reagan Administration allowed this plague to take its course from 1981 to 1985. Not until it was realized that AIDS would not stay confined to the designated deviant population was legitimized political action taken to contain the virus and control its symptoms.

Engineering the death of populations by neglect is not a recent innovation. Certainly the Irish genocide at the time of the potato blight indicates that this strategy has been around for while (although it should be remembered that this genocide was also primarily done for land and resources, and less for reasons of purity). Death by neglect is a haunting reminder that Social Darwinism and the anti-welfare recommendations of Malthus and Spencer in the time of early capital are not only alive and well, but are once again gaining in strength.

For acts of passive genocide to be perceived as legitimate (natural), the public must participate in eugenic ideology. It must believe that the species is in a biological process that is striving for perfection through a selection process. It must believe that some populations are more fit than others. It must desire to emulate the fit, and to have faith that the unfit will be eliminated. With this belief in place, social vectors of power only have to contextualize the ideological system in a particular social moment to see its design come to fruition for a political-economy that is encoded directly into the flesh. Returning to the announcement from Ars Electronica, an indicator of this process at work can be observed when we read: "Complex tools and technologies are an integral part of our evolutionary 'fitness.' Genes that are not able to cope with this reality will not survive the next millennium." Who benefits from beliefs such as this? Those who profit most from the development of technocratic pancapitalism. There is not a shred of evidence that nature selects for genes with a predisposition for using complex tools. In fact, if survival is taken as the signifier of fitness, those who use complex tools are a small and stable minority of the world's population, which would indicate that they are less fit. The majority and expanding populations do not use complex tools. (What is truly odd is that such rhetoric implies that "quality of life" is a characteristic that demonstrates fitness and adaptability. This is a peculiar return to the Calvinist belief in finding signs that one is in God's grace by one's proximity to economic bounty). It seems just as likely that complex tools are signs of devolution, or even the source of species destruction. What is clear is that the power vectors which currently engineer social policy are at the moment selecting for and rewarding those who can use complex tools and punishing those who cannot, and that this intentional process is at times passed off as a natural development.

The sweeping condemnation of those outside technoculture bodes badly for less technologically saturated societies, since they presently appear to be "unfit" according to this line of thinking. Traces of the colonial narrative replay themselves in this rhetoric, since technoculture is not accessible to the grand majority of nonwestern races and ethnicities. At the same time, the colonial narrative is being reconfigured for postwar technoculture. As women are brought into the bureaucratic and technocratic workforce, fitness designated by biological characteristics is starting to be replaced by fitness designated by behavior. This way, power vectors have an alibi which masks the traces of the colonial narrative alive in technoculture, but which can also allow them to embrace "fit" individuals that emerge from "unfit" populations.

Conclusion

Two key problems occur in attempting to use evolutionary theory in the analysis of cultural development. First, presenting cultural development as analogous to biological development is like trying to hammer a square peg into a round hole. There is little basis for likening a blind, groping process of species configuration within a chaotic, uncontrolled environment to a rationally engineered process of social and economic development within an orderly, controlled environment. Retrograde notions of cultural development, such as providence, progress, and manifest destiny, have more explanatory power, because they at least recognize intentional design in cultural dynamics, and at the very least they imply the existence of a power structure within the cultural environment. Evolutionary theory, in its social sense, is blind to the variable of power, let alone to the inequalities in its distribution.

The second problem is historical. Since the application of evolutionary theory has continuously been the foundational rhetoric and justification of social atrocity for the last 150 years, why would anyone want to open this Pandora's box yet again? At a time when biotech products and services are being developed that will allow imperatives of political economy to be inscribed directly into the flesh and into its reproductive cycle, why would anyone want to use a theoretical system with little, if any, informative power, that if deployed through pancapitalist media filters will promote eugenic ideology? While it cannot be denied that all inquiries for the purpose of gaining knowledge bring with them a high probability that the information collected could be misused in its application, in the case of social evolutionary theory, the historical evidence that it will be misused is overwhelming. This situation is not fuzzy enough to make this roll of the dice a smart gambit, and the good intentions of individuals who engage this discourse will not save it from capitalist appropriation and reconfiguration to better serve its authoritarian and nihilistic tendencies.

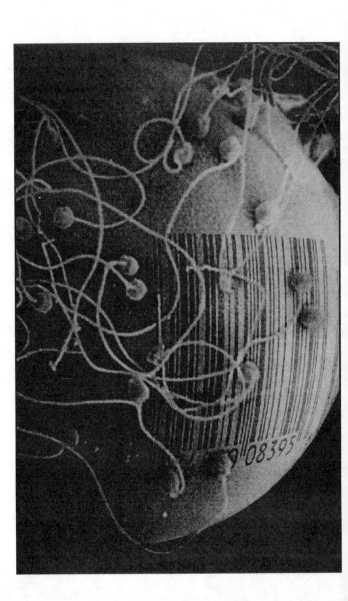

3

The Coming of Age
of the Flesh Machine*

Over the past century, the two machines that comprise the general state apparatus have reached a level of sophistication which neither is likely to transcend. These complex mechanisms, the war machine and the sight machine, will go through many generations of refinement in the years to come; for the time being, however, the boundaries of their influence have stabilized.

The war machine is the apparatus of violence engineered to maintain the social, political, and economic relationships that support its continued existence in the world. The war machine consumes the assets of the world in

---

*This article was originally published in *Electronic Culture*, Tim Druckrey, ed., New York: Aperture, 1996.

classified rituals of uselessness (for example, missile systems that are designed never to be used, but rather to pull competing systems of violence into high-velocity cycles of war-tech production) and in spectacles of hopeless massacre (such as the Persian Gulf war). The history of the war machine has generally been perceived in the West as history itself (although some resistance to this belief began during the 19th century), and while the war machine has not followed a unilinear course of progress, due to disruptions by moments of inertia caused by natural disasters or cultural exhaustion, its engines have continued to creep toward realizing the historical construction of becoming the totality of social existence. Now it has reached an unsurpassable peak—a violence of such intensity that species annihilation is not only possible, but probable. Under these militarized conditions, the human condition becomes one of continuous alarm and preparation for the final moment of collective mortality.

The well-known counterpart of the war machine is the sight machine. It has two purposes: to mark the space of violent spectacle and sacrifice, and to control the symbolic order. The first task is accomplished through surveying and mapping all varieties of space, from the geographic to the social. Through the development of satellite-based imaging technologies, in combination with computer networks capable of sorting, storing, and retrieving vast amounts of visual information, a wholistic representation has been constructed of the social, political, economic, and geographical landscape(s) that allows for near-perfect surveillance of all areas, from the micro to the macro. Through such visualization techniques, any situation or population deemed unsuitable for perpetuating the war machine can be targeted for sacrifice or for containment.

The second function of the sight machine, to control the symbolic order, means that the sight machine must generate representations that normalize the state of war in everyday life, and which socialize new generations of individuals into their machinic roles and identities. These representations are produced using all types of imaging technologies, from those as low-tech as a paint brush to ones as high-tech as supercomputers. The images are then distributed through the mass media in a ceaseless barrage of visual stimulation. To make sure that an individual cannot escape the imperatives of the sight machine for a single waking moment, ideological signatures are also deployed through the design and engineering of all artifacts and architectures. This latter strategy is ancient in its origins, but combined with the mass media's velocity and its absence of spatial restrictions, the sight machine now has the power to systematically encompass the globe in its spectacle. This is not to say that the world will be homogenized in any specific sense. The machinic sensibility understands that differentiation is both useful and necessary. However, the world will be homogenized in a general sense. Now that the machines are globally and specifically interlinked with the ideology and practices of pancapitalism, we can be certain that a hyper-rationalized cycle of production and consumption, under the authority of nomadic corporate-military control, will become the guiding dynamic of the day. How a given population or territory arrives at this principle is open to negotiation, and is measured by the extent to which profit (tribute paid to the war machine) increases within a given area or among a given population.

In spite of the great maturity of these machines, a necessary element still seems to be missing. While representation has been globally and rationally encoded with the imperatives of pancapitalism, the flesh upon which these codings are further inscribed has been left to reproduce and develop in a less than instrumental manner. To be sure, the flesh machine has intersected both the sight and war machines since ancient times, but comparatively speaking, the flesh machine is truly the slowest to develop. This is particularly true in the West, where practices in health and medicine, genetic engineering, or recombinant organisms have thoroughly intersected nonrational practices (particularly those of the spirit). Even when they were secularized after the Renaissance, these practices have consistently been less successful, when compared to their counterparts, in insuring the continuance of a given regime of state power. Unlike the war machine and the sight machine, which have accomplished their supreme tasks—the potential for species annihilation for the former, and global mapping and mass distribution of ideologically coded representation for the latter—the flesh machine has utterly failed to concretize its imagined world of global eugenics.

The simple explanation for the flesh machine's startling lack of development is cultural lag. As the West shifted from a feudal to a capitalist economy, demonstrating the benefits of rationalizing production in regard to war was a relatively simple task. National wealth and border expansion were clearly marked and blended well with the trace leftovers of feudal ideology. Manifest destiny, for example, did not stand in contradiction to Christian expansionism. War, economy, politics, and ideology (the slowest of social manifestations to change) were still working toward a common end (total domination). The rationalization of the flesh, however, could not find a point of connection with theologically informed ideology. Flesh ideology could only coexist as parallel rather than as intersecting tracks. For this reason it is no surprise that one of the fathers of flesh machine ideology was a man of God. The work of Thomas Malthus represents the ideological dilemma presented to the flesh machine on the cusp of the feudal/capitalist economic shift.

Malthus argued that the flesh did not have to be rationalized through secular engineering, since it was already rationalized by the divine order of the cosmos designed by God Himself. Although the nonrational motivation of original sin would guarantee replication of the work force, God had placed "natural checks" on the population, so only those who were needed would be produced. The uncivilized lower classes could be encouraged to have as many children as possible without fear that the population would overrun those in God's grace, because God would sort the good from the bad through famine, disease, and other natural catastrophes. For this reason, the flesh could be left to its own means, free of human intervention, and human progress could focus on fruition through economic progress. Spencerian philosophy, arriving half a century later, complemented this notion by suggesting that those fit for survival would be naturally selected in the social realm. The most skillful, intelligent, beautiful, athletic, etc., would be naturally selected by the structure of the society itself—that of "open" capitalist competition. Hence the flesh machine was still in no need of vigorous attention; however, Spencer did act as a hinge for the development of eugenic consciousness. Spencer constructed an ideological predisposition for conflating natural and social models of selection (the former arrived a decade or so after Spencer's primary theses were published). This made it possible for genetic engineering to become a naturalized social function, intimately tied to social progress without being a perversion of nature—in fact, it was now a part of nature. At this point eugenic consciousness could continue to develop uninterrupted by feudal religious dogma until its traces evaporated out of capitalist economy, or until it could be better reconfigured to suit the needs of capitalism. While the idea of a eugenic world continued to flourish in all capitalist countries, and culminated in the Nazi flesh experiment of the 30s and early 40s, the research never materialized that would be necessary to elevate the flesh machine to a developmental level on a par with the war machine.

Perhaps there is an even simpler explanation. Machinic development can only occur at the pace of one machine at a time, since scarce resources allow for only so much indirect military research. After the war machine came to full fruition with the implementation of fully matured total war during World War II, along with the attendant economic expansion, it became possible to allocate a generous helping of excess capital for the expansion of the next machine. In this case, it was the sight machine which had proved its value during the war effort with the development of radar and sonar, and thereby jumped to the front of the line for maximum investment. It was also clearly understood at this point that global warfare required new attention to logistic organization. The road between strategic and tactical weapons and logistical needs had leveled out, and this realization also pushed the sight machine to the front of the funding line. Conversely, the need of the Allied powers to separate themselves ideologically as far as possible from Nazi ideology pushed the desired development of the flesh machine back into the realm of nonhuman intervention. Consequently, the alliance between the war machine and the sight machine continued without interruption, delivering ever-increasingly sophisticated weapons of mass destruction. It also created an ever more enveloping visual/information apparatus—most notably satellite technology, television, video, computers, and the Net.

While the war machine reached relative completion in the 60s, the sight machine did not reach relative completion until the 80s (die-hard Web-users might want to argue for the 90s). Now a third machine can claim its share of excess capital, so the funds are flowing in increasing abundance to a long deferred dream. The flesh machine is here. It has been turned on, and like its siblings, the war machine and the sight machine, it cannot be turned off. As is to be expected, the flesh machine replicates elements of the sight and war machines in its construction. It is these moments of replication which are of interest in this essay.

A Brief Note on Scientific Imagination, Ethics, and the Flesh Machine

In the best of all possible worlds, ethical positions relevant to the flesh machine would be primary to any discussion about it. In fact, to read the literature on the flesh machine (which at this point is dominated by the medical and scientific establishments), one would think that ethics is of key concern to those in the midst of flesh machine development; however, nothing could be further from reality. The scientific establishment has long since demonstrated that when it comes to machinic development, ethics has no real place other than its ideological role as spectacle. Ethical discourse is not a point of blockage in regard to machinic development. Take the case of nuclear weapons development. The ethical argument that species annihilation is an unacceptable direction for scientific inquiry should certainly have been enough to block the production of such weaponry; however, the needs of the war machine rendered this discourse silent. In fact, the need of the war machine to overcome competing machinic systems moved nuclear weapons development along at top velocity. Handsome rewards and honors were paid to individuals and institutions participating in the nuclear initiative. In a word, ethical discourse was totally ignored. If big science can ignore nuclear holocaust and species annihilation, it seems very safe to assume that concerns about eugenics or any of the other possible flesh catastrophes are not going to be very meaningful in its deliberations about flesh machine policy and practice. Without question, it is in the interest of pancapitalism to rationalize the flesh, and consequently it is in the financial interest of big science to see that this desire manifests itself in the world.

Another problem with machinic development could be the institutionally-contained panglossian reification of the scientific imagination. Consider the following quote from Eli Friedman, president of International Society for Artifial Organs, in regard to the development of artificial organs:

> Each of us attempting to advance medical science—whether an engineer, chemist, theoretician, or physician—depends on personal enthusiasm to sustain our work. Optimistic, self-driven investigators succeed beyond the point where the pessimist, convinced that the project cannot be done, has given up. Commitment to the design, construction, and implanting of artificial internal organs requires a positive, romantic, and unrestrained view of what may be attainable. Members of our society share a bond gained by the belief that fantasy can be transformed into reality.

and:

> ISAO convenes an extraordinary admixture of mavericks, "marchers to different drums," and very smart scientists capable of converting "what if" into "why not."

These lovely rhetorical flourishes primarily function to rally the troops in what will be a hard-fought battle for funding. It's time to move fast (the less reflection the better) if the AO model is to dominate the market; after all, there is serious competition from those who believe that harvesting organs from animals (transgenic animals if need be) is the better path along which to proceed. But it is the subtext of such thinking that is really of the greatest interest. From this perspective, science lives in a transcendental world beyond the social relationships of domination. If something is perceived as good in the lab, it will be good in the world, and the way a scientist imagines a concept/application to function in the world is the way it will in fact function. The most horrifying notion, however, is the idea (bred from a maniacal sense of entitlement) that "if you can imagine it, you may as well do it," as if science is unconnected to any social structures or dynamics other than utopia and progress.

Perhaps the only hope is that the funding and the optimism becomes so excessive that it undermines machinic development. Star Wars is a perfect example of incidental resistance from the scientific establishment. During the Reagan-era big bonanza for war machine funding, the most ludicrous promises were made by big science in order to obtain research funds. The result was a series of contraptions that truly defines the comedy of science. Two of the finest examples are the rail gun that self-destructed upon launching its pellet projectile, and the deadly laser ray that had a range of only three feet. While the American taxpayers might see red over the excessive waste, a major section of the scientific establishment was apparently distracted enough by the blizzard of money that they failed to make any useful lethal devices.

If I Can See It, It's Already Dead

The war machine and the sight machine intersect at two key points—in the visual targeting of enemy forces (military

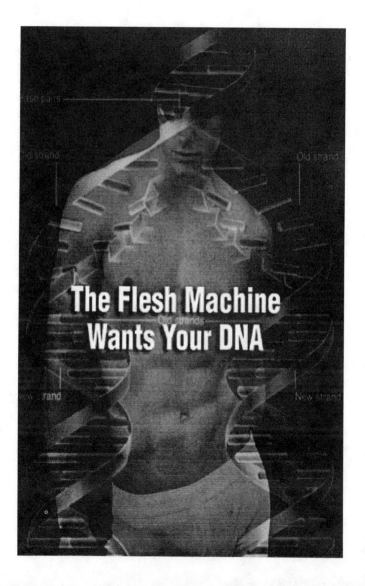

The Flesh Machine
Wants Your DNA

sites, production sites, and population centers), and in visualizing logistical routes. Once sited and accurately placed within a detailed spatial grid, the enemy may be dispatched at the attacker's leisure, using the most efficient routes and means of attack. As long as the enemy can remain invisible, determining proper strategic action is difficult, if not impossible. Hence any successful offensive military action begins with visualization and representation. A strong defensive posture also requires proper visual intelligence. The better the vision, the more time available to configure a counterattack. The significant principle here—the one being replicated in the development of the flesh machine—is that vision equals control. Therefore the flesh machine, like its counterparts, is becoming increasingly photocentric.

Not surprisingly, much of the funding for the flesh machine is intended to develop maps of the body and to design imaging systems that will expedite this process. From the macro to the micro (the Human Genome Project being the best known), no stone can remain unturned. Every aspect of the body must be open to the vision of medical and scientific authority. Once the body is thoroughly mapped and its "mechanistic" splendor revealed, any body invader (organic or otherwise) can be eliminated, and the future of that body can be accurately predicted. While such developments sound like a boon to humanity, one need not be an expert in the field to be skeptical of such prospects.

While it is hard to doubt the success of the war machine in reducing military activity to the mechanized (that is, fully rationalized structures and dynamics), it is questionable whether the body can be reduced to a similar state regardless of how well it is represented. One major problem is that the body cannot be separated from its environment, since so many of its processes are set in motion by environmental conditions. For example, a toxic environment can produce undesirable effects in the body. Visual representation alerts medicine to an invasion, so action can be taken to contain or eliminate the invader. In this situation, medicine is reactive rather than preventive, and treats only the effect and not the cause. In fact, it diverts causality away from ecological pathologies, and reinvests it back in the body. In this manner, medicine becomes an alibi for whatever created the toxic situation that infected the body in the first place, by acting as if the infectant emerged internally. The problem raised here is the limited frame of representation in regard to the body map, in conjunction with an emphasis on tactical solutions to physical pathologies. This situation is, of course, understandable, since strategic action would have an undermining effect on the medical market. The one exception to this rule is when the toxic body emerges due to behavioral factors. In this case, the scientific/medical establishment can expand its authority over the body by suggesting and often enforcing behavioral restrictions on patients. In this situation, the science and medical establishment functions as a benevolent police force deployed against individuals to better mold them to the needs of the state.

To complicate matters further, flesh machine science and medicine have the unfortunate but necessary habit of putting the cart before the horse. The flesh machine, unlike its counterparts, does not have the luxury of developing its visual and weapons systems simultaneously, nor can weapons development precede advanced visual capabilities. The visual apparatus must come first. For example, antibiotics probably could not have been invented before the development of a microscope. Consequently, as in most research and development, a shotgun method is employed, whereby all varieties of vision machines are developed in the hopes that a few may be of some use. This leads to thrilling headlines like the following from Daniel Haney of the Associated Press: "Brain Imagery Exposes a Killer." What this headline refers to is a new medical map, acquired through the use of positron-emission tomography, which reveals the part of the brain affected by Alzheimer's disease, and the degree to which the brain has been eroded by the disease. This map can help physicians to diagnose Alzheimer's up to ten years before symptom onset. The comedy begins with the admission that there is no way to predict when symptoms will begin to appear, and that there is still no known treatment for the disease. All that medical science can do is tell the patient that s/he has the disease, and that s/he will be feeling its effect sometime in the future. The excitement over being able to visualize this disease comes from the belief that if the disease can be seen, then cure is near at hand. Or, in the words of the war machine, "If I can see it, it's already dead."

Since the process of visualization and representation in this case is at best only an indication of a far-off possibility for cure, and hence is of little use for the patient already diagnosed with the disease, it must be asked: Who could benefit from this information? Alzheimer's is in fact doubly problematic because it can be visualized before symptom onset, and because genetic mapping can also be used to indicate an individual's likelihood of developing it. The flesh machine's intersection with the surveying function of the sight machine becomes dramatically clear in this situation. Those who would benefit most from this information are insurance companies and the employer of the person likely to be afflicted with the ailment. Such information would be a tremendous cost-cutting device for both. However, ethical discussions about collecting biodata lead one to believe that such information would remain confidential in the doctor-patient relationship. Perhaps privacy will be maintained. However, it seems more likely that if the information is perceived to lead to significantly higher profits, resources will be allocated by corporate sources to acquire it. The most common strategy to watch for is legislative initiatives pursued under the spectacle of benevolence. Mandatory drug testing for some private and public employment, under the authority of employee and public security, is an example of the means by which privacy can be eroded.

Finally there is the problem of representation itself. As the war machine demonstrates, the greater the visualization of a frontier territory, the greater the degree of contestation at the visualized sight. In other words, the more that is seen, the more power realizes what needs to be controlled and how to control it. The brain is certainly going to be the key, but happily, at this point, the research is too immature to warrant strategic intervention on behalf of state power. There are, however, good indicators of how the coming battle will take shape. One need only think of the visualization of the body and its connection to varieties of smoking bans from the legalistic to the normative, or in terms of populist countersurveillance, the relationship of toxins (DDT, for example) in the environment to body visualization, to understand the connection between vision, discipline, and contestation. The prizewinner, however, is the visualization of uterine space. Feminist critics have long shown how this point of ultra-violent contestation is but the beginning of the age of flesh machine violence. (This is also a point of great hope, as the discourse of the flesh machine has been appropriated from the experts. At the same time, this conflict has shown how fascist popular fronts are just as adept at appropriation). In regard to uterine space, feminist critics have consistently pointed out that this variety of representation loads the ideological dice by presenting the space as separate from the wholistic bio-system of the woman, thus reinforcing the notion of "fetal space." This idea acts as a basis for "fetal rights," which are then argued as taking precedence over the rights of women.

A new era of bio-marginality has surely begun. Certainly this situation will only be reinforced by the visualization of either diseases or abnormalities (actual or potential) in subjects soon to be classified under the sign of the unfit. The unfit will be defined in accordance with their utility in relationship to the machine world of pancapitalism. The mapped body is the quantified body. Its use is measured down to the penny. Without such a development, how could any consumer trust in the markets of the flesh machine?

Selling Flesh

One of the oldest manifestations of the flesh machine is the idea of engineering the breeding of plants and livestock to produce what are perceived to be the most functional products within a given cultural situation. Increased knowledge about this task has certainly contributed to the great abundance in the food supply in the first world, thus shifting an individual's relationship to food from one of need to one of desire. In light of this achievement, industrial food producers have been faced with the task of developing foods that meet the logistical demands of broad-based distribution, while still maintaining a product that the manufacturer can market as desirable. The most productive solution thus far is the manufacture of processed foods; however, the market for food cannot be limited to processed food. The desire for perishable foods is too deeply etched into the culture, and no amount of spectacle can root out this desire. Fortunately for the producers of perishable foods, the product and the market can be rationalized to a great extent. This particular market is of interest because it provides at this moment the best illustration of the market imperatives that are being replicated in the industrial production and distribution of human flesh products. (This is not to say that flesh production will not one day be more akin to processed food, it is only to argue that at present the means of production are still too immature).

To better illuminate this point, consider the case of apples. At the turn of the century, there were dozens of various types of apples available to the buying public. Now when a consumer cruises through a supermarket in search of apples, the choice has been limited to three (red, green, and yellow). Choice has become increasingly limited partly because of logistical considerations. Like most perishable fruits and vegetables these days, apples are bred to have a long shelf life. In order to have apples all year round, they must be transported from locations that have the conditions to produce them when other locations cannot. Hence these apples must be able to survive an extended distribution process, and not all varieties of apples are capable of resisting rotting for long periods of time. However, logistics alone does not adequately explain choice limitations. Perhaps more important to the formula are market considerations.

Marketing agencies have understood for decades that desire is intensified most through visual appeal. How a product looks determines the probability of a consumer purchase more than any other variable. For apples, the consumer wants brightly colored surfaces, a rounded form, and white inner flesh. In other words, consumers want the perfect storybook apple that they have seen represented since they were children. Apples are bred to suit the cultural construction of "an apple," and only a few varieties of apples can simulate this appearance and meet this desire. This situation is yet another case of Baudrillard's universe of platonic madness, where consumers are caught in the tyranny of representation that passes as essence.

Along with the domination of vision, there comes the need of the producer to offer the consumer a reliable product, meaning that the apples one buys tomorrow will look and taste like the ones bought today. Consequently, there is an elimination of sense data other than the visual. If all that is needed to excite desire is a good visual, why bother to develop taste and smell? Especially when a good product can be guaranteed if it is completely tasteless (one can be sure that the apple purchased tomorrow will taste like the one purchased today)? In this situation, the tyranny of the image becomes glaringly apparent; one would think that smell and taste would be the dominating senses when buying foods, since they would best articulate the pleasure of consumption. Not so, it is vision, and unfortunately many of the most tasty apples do not look very good because they have none of the necessary storybook appeal. Consequently various types of apples have been eliminated, or limited to distribution in localized markets.

If the principles of product reliability and visual appeal are applied to the production/consumption components (as opposed to those concerned with control) of the flesh machine, the reasons for some recent developments become a little clearer. The first problem that flesh producers must face is how to get a reliable product. At present too little is known about genetic processes to fulfill this necessary market imperative. Consequently, they have had to rely on fooling the naive consumer. For example, one characteristic commonly sought after by those in the techno-baby market is intelligence. Unfortunately this characteristic cannot be guaranteed; in fact, flesh producers haven't the slightest idea how to replicate intelligence. However, they can promise breeding materials from intelligent donors. While using the sperm of a Nobel Prize winner in no way guarantees a smart child, and doesn't even increase the probability (nor does it decrease the probability of having a below average child), flesh dealers are able to use false analogies to sell their product. (If two tall parents have a child, the probability of the child being tall is increased, so wouldn't it be correct to say that if two people of above average intelligence have a child, that it would increase the probability that the child will have above average intelligence?) Many consumers believe this line of thought (the myth of hard genetic determinism has always been very seductive) and are therefore willing to pay higher prices for the sperm of an intelligent man than they are for the sperm of an average donor. Although this fraud will probably not continue indefinitely in the future, an important ideological seed is being sown. People are being taught to think eugenically. The perception is growing that in order to give a child every possible benefit in life, its conception should be engineered.

Another common strategy to better regulate flesh products is to take a genetic reading of the embryo while still in the petri dish. If a genetic characteristic is discovered that is deemed defective, the creature can be terminated before implantation. Again, parents-to-be can have their eugenic dreams come true within the limits of the genetic test. Even parents using the old-fashioned method of conception at least have the option of visualization (sonar) to make sure that the desired gender characteristic is realized. In each of these cases, better visualization and representation, along with an expanded range of genetic tests, will help to insure that desired characteristics are always a part of the flesh product, which leads to the conclusion that better vision machines are as important for profit as they are for control.

At the same time, remember that the marketing practices of postmodernity do not wholly apply to the flesh machine, and at present tend to function on an as-needed basis. Fertility clinics, for example, participate as much in the economy of scarcity (although it must be noted that these products and processes do not intersect the economy of need) as they do in the economy of desire. While they may use the practices described above, they also have the luxury of being the only option for those who have been denied the ability to produce flesh materials. Those clinics that can boast a product success rate of over 20% (most notably the Center for Reproductive Medicines and Infertility at New York Hospital-Cornell Medical Center, with a success rate of 34%) cannot meet the demand for their goods and services. Apparently, the market for flesh goods and services has been preconstructed in the bio-ideology of capitalism.

When Worlds Collide

Assuming that the flesh machine is guided by the pancapitalist imperatives of control and profit, what will occur if these two principles come into conflict with one another? This has been known to happen as social machines march toward maturity. The sight machine is currently facing this very contradiction in the development of the Net. Currently the Net has some space that is relatively open to the virtual public. In these free zones, one can get information on anything, from radical politics to the latest in commodity development. As to be expected, a lot of information floating about is resistant to the causes and imperatives of pancapitalism, and from the perspective of the state is badly in need of censorship. However, the enforcement of limited speech on the Net would require measures that would be devastating to on-line services and phone service providers, and could seriously damage the market potential of this new tool. (The Net has an unbelievably high concentration of wealthy literate consumers. It's a market pool that corporate authority does not want to annoy). The dominant choice at present is to let the disorder of the Net continue until the market mechanisms are fully in place, and the virtual public is socialized to their use; then more repressive measures may be considered. Social conservatism taking a back seat to fiscal conservatism seems fairly representative of pancapitalist conflict resolution. The question is, will this policy replicate itself in the flesh machine?

A good example to speculate on in regard to this issue is the ever-elusive "gay gene," always on the verge of discovery, isolation, and visualization. Many actually anxiously await this discovery to prove once and for all that gayness is an essential quality and not just a "lifestyle" choice. However, once placed in the eugenic matrix this discovery might elicit some less positive associations. In the typical alarmist view, if the gene comes under the control of the flesh machine, then it will be eliminated from the gene pool, thus giving compulsory heterosexuality a whole new meaning. Under the imperative of control this possibility seems likely; however, when the imperative of marketability is considered, a different scenario emerges. There may well be a sizable market population for whom the selection of a gay gene would be desirable. Why would a good capitalist turn his back on a population that represents so much profit, not to mention that gay individuals as a submarket (CAE is assuming that some heterosexuals would select the gay gene too) must submit to the flesh machine to reproduce? Again, market and social imperatives come into conflict, but it is unknown which imperative will be selected for enforcement.

Such an issue at least demonstrates the complexity of the flesh machine, and how difficult the task of analyzing this third leviathan will be. What is certain is that the flesh machine is interdependent with and interrelated to the war machine and the sight machine of pancapitalism, and that it is certainly going to intensify the violence and the repression of its predecessors through the rationalization of the final component (i.e., the flesh) of the production/ consumption process. Until maps are produced for the purpose of resistance and are crossed-referenced through the perspectives of numerous contestational voices, there will be no way, practical or strategic, to resist this new attack on liberationist visions, discourse, and practice.

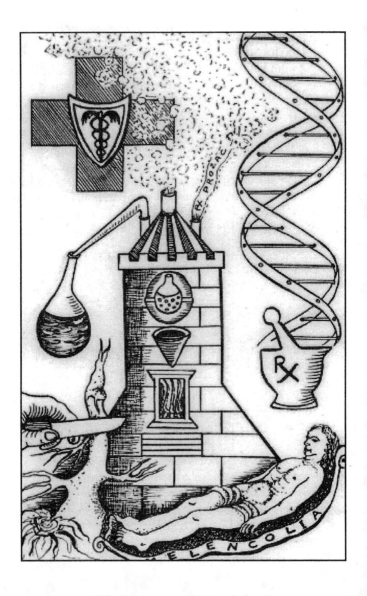

4

Buying Time
for the Flesh Machine:
Pharmacology and Social Order

Drugs are a part of everyday life in pancapitalist society, and serve
a variety of social purposes including the medicinal, the
recreational, and upon occasion the spiritual. These categories overlap and intersect to a greater or lesser degree,
depending on the context in which they are used. Nestled
within this pharmacological collection is another category
of drugs that also intersects the other varieties. These drugs
are designed to "normalize" behavior as well as the social
presentation of the body. They are usually perceived under
the sign of the medicinal, but they are different in quality
and function. Unlike drugs used to prolong life, where
their quality is primarily measured by a patient's proximity
to death (ideological factors are a secondary measure-
ment), the quality of drugs used to normalize behavior is
measured by the patient's willingness to conform to social

imperatives and the patient's ability to integrate he/rself into specific social contexts beneficial to the given political economy. Within the context of pancapitalism, the general social imperatives inscribed on the individual are those of production, consumption, and order under the guiding metaprinciple of efficiency. When individuals break from these imperatives by taking up disruptive behavior that is beyond the norm, but is still perceived as manageable, biochemical intervention becomes a viable, simple, and profitable option for ideological re-inscription. The drugs used for this purpose primarily function as a means of social control, and they maintain a social environment that is valued above and beyond the physical, mental, or spiritual well-being of the individual.

The problem with such biochemical interventions is that they are not a very efficient means of ideological re-inscription. Too often, the drugs used have side effects that are as counterproductive as the behavior they seek to eliminate. This situation may change as researchers learn more about biochemistry, but biochemical intervention is not the most desirable from the perspective of pancapitalism. Interventionist drugs, being merely symptom managers, do not function preventively, and their great fiscal value is offset by their modest value as behavioral control mechanisms. The answer to this problem is to eliminate any biological cause for "anti-social" behavior by ideologically designing and engineering the flesh. This goal is among the key mandates of the flesh machine. Unfortunately, the flesh machine is at present too immature to implement practical strategies in this biological arena, and hence cannot meet this goal. Until it can prepare strategies of flesh intervention that reduce deviant behavior, social control drugs can be used temporarily to reinforce the prison walls of social imperatives and normative behavior. This is not to say that drugs for the purpose of social control will completely disappear when the flesh machine matures; it is only to say that biochemical intervention will be minimized in the future. Social control drugs will always have a place in reforming or eliminating deviant behavior that is primarily caused by cultural conditions, and to enhance normative behavior (smart drugs).

Neutralizing Emotion

Most antidepressants and mood stabilizers are designed for people who are still socially functional and reasonably well integrated into their social systems. The problem with these people is that their mood swings can disrupt the spaces of production and consumption. Efficient work requires stabilized, rational, instrumental behavior. Behaviors which fall outside of these parameters are considered undesirable and disruptive. Excessive emotion that does not originate within the process of work itself decreases a worker's output. This would not be so bad if it were self-contained; unfortunately, work is generally a group process. When excessive emotion affects a worker so much that it manifests itself in behavior, it initiates a social current that has a detrimental effect on the other workers, both in terms of morale and behavior. Consequently, interventionist practices become necessary. Antidepressants and mood stabilizers arrest the behavioral symptoms of excessive emotion. While they do not necessarily help the individual using them find peace of mind, they do tend to function well by stopping the individual from becoming a catalyst for detrimental social currents in the space of production.[1]

The downside of this strategy is that chemical intervention of this class cannot be too widespread, since the manufacture of desire requires an affective response from its target populations. How could store displays, impulse racks, product advertisements, political advertisements. architectural designs, and other manifestations of manufactured desire possibly function if the population were emotionally neutral? Unlike the environment of production, where stable, rational, and instrumental action is required, the environment of excess consumption requires unstable, nonrational, and affective action. Without it, consumers could not be convinced to buy that which they do not really need (preferably with the money that they have yet to earn). The product providers must establish an emotional-based pleasure switch in consumers that can be activated by spectacular means. Mood-stabilizing drugs confound the strategies to do this, so—ironically enough— movement away from such chemicals is also desirable, when placed in the context of consumption. Mood stabilizers place capital in the awkward position of using a social control strategy that negates itself as it moves from the context of production to consumption.

Children's Pathologies

Two common pathologies believed to cause behavior problems in children are Attention Deficit Disorder (ADD) and Hyperactivity. The former has recently been recognized as an affliction which also affects adults, but medical interest in the disorder is still centered on children. One must wonder why it is so important to diagnose and manage this ailment in children, and why it is less important to do so in adults. The answer is primarily cultural. Once children leave the confines of domestic space, and are shuttled off to their first institutional environment (school), a new level of socialization begins. The education system accelerates the process of teaching the child that there is a hierarchical social order, that it is meticulously rule-laden, and that these rules must be followed. When it functions as intended, mass socialization teaches the child obedience to authority, acceptance of the instrumental fragmentation of time, the importance of repetitive labor, and a tolerance for boredom simply by repeatedly placing the child within the bureaucratic context day after day. These social fundamentals set the context in which the child learns verbal and analytic skills. If this process is successfully completed, the young adult will be adequately prepared to begin work as a low-level bureaucrat, or as a semi-skilled or skilled laborer. Others who excel in verbal or analytic skills can move on to further education and train for a place in managerial, professional, or high-level technocratic work. Those who are for any reason unable to cope with any part of the system are cast adrift.

The process of forcing out the incapable (whether the incapacity is due to physical, psychological, sociological, or economic reasons) takes time. Until it becomes clear which students are unable to "adapt" to the process, all are served to insure that both the work force itself and the ideological conditions for an efficient work force are continually replicated. Given the significance of this general process, one which the adult transcends by entering a more specialized field, it is little wonder that behavior which disrupts the replication of the primary educational imperatives will not be tolerated. Children must know that they must be at certain places at certain times engaging in specific tasks. They must know that they must focus their attention on work, no matter how boring, when they are told to do so. They must know that they must recognize and obey those who are of a higher rank. And if they do not, bureaucratic officials can only conclude that they are mentally ill or incapable (resistance to instrumental consciousness is rarely viewed as a sign of intelligence). For those who are found to be mentally ill (in this case suffering from ADD or hyperactivity), biochemical intervention is a necessity. Like the worker on antidepressants, the child treated with drugs can better integrate he/rself into the social environment, but these drugs will ensure that the teacher's work is not interrupted by the child, and that the other children's socialization process is not disrupted. After all, what use is a child who resists the given social order? What need is there for a child who would rather drift into the realm of the imagination instead of doing he/r boring assignment, or a child who would rather engage the emergent possibilities of play than engage the overdetermined structure of work? Such children are not only disruptive, but potentially serve as damaging role models for other children, since they represent the beginnings of a formation of resistant thinking and activity. Unfortunately, educators who see a significant place for such children and reject interventionist drug practices geared toward maintaining the status quo of the bureaucratic environment are marginalized as radical pedagogues.

An ailment like ADD, or any other anti-work disorder, is seldom diagnosed in the adult environment, because it is not needed. Since it is assumed that an adult's socialization process is complete, and hence the punishment for poor production is known (dismissal from duty), the option of biochemical intervention need only be suggested. Employers expect that adults will acquiesce out of conditioned fear. However, if the socialization process has really worked effectively, adults will volunteer for intervention as soon as they realize that production is dropping.

Relieving Anxiety and Stress

Like antidepressants, anxiety and stress relievers are designed primarily for those who are still functioning adequately in their social environments. In fact, for these drugs to be profitable, individuals must be competent enough to realize that their mental/physical condition is not suitable for a particular social environment, and at least have modest knowledge of how and where to "get help." Since anxiety and stress disorders are so common in everyday life, considerable research and development has been done to improve the products for this mass market. The members of this class of interventionist drugs have been refined to minimize addictive characteristics, and to eliminate recreational and other noninstrumental characteristics. Unlike in the 50s and 60s when powerful and potentially addicting drugs, like Seconol, could be prescribed for various stress disorders, physicians now prescribe less powerful drugs such as Ativan or Xanax. These drugs have less euphoric or stupefying effects than their predecessors while maintaining the desired effects on the individual, and hence are perceived as improvements over their predecessors. In cases where drugs could not be cleansed of their recreational characteristics, nor be redeployed in specialized medical markets, they were simply made illegal (methaqualone for example). The result is an array of products that will maintain the mass market for stress relief, while minimizing unacceptable side effects.

Indeed such a market for pharmaceuticals is currently more profitable than ever. In the US, as work intensifies, the workday grows longer, and real wages dramatically decrease, stress levels rise. The stress effect is doubled in the economy of excess: Individuals not only worry about survival, but are taught to fear losing the goods which they have accumulated. The pancapitalist medical establishment has responded to this social development by offering their services as mental mechanics to keep unstable bodies performing at peak levels. Managed drug intervention is a primary tactic used to accomplish this task. Even corporations have begun to respond by offering their employees access to exercise classes and gymnasiums, and a rare few offer access to on-the-clock tactical meditation, hypnosis, and rest. The important factor here is that the body is perceived as the culprit in need of normalization, because it is failing to adequately adapt to its environment. However, since the environment in which pathologically stressful conditions occur is intentionally designed to exceed human potential, the real culprit causing bodily disruption is the social situation in which the individual finds he/rself trapped. As to be expected, few attempts are made to correct the pathological conditions which cause individuals to work themselves into sickness, and thereby place themselves in need of drug intervention. Quite the opposite—environmental conditions in the space of production and service are becoming more detrimental to human well-being, and the responsibility for maintaining one's body in this hostile environment is being placed squarely on the individual.

Here the mandate of the flesh machine is to engineer the body in a manner that increases its ability to more readily adapt to cultural conditions that have transcended human biology. On the one hand, the flesh machine must select characteristics that will increase production/service, such as increased concentration and stamina, while on the other hand, it must eliminate stress mechanisms (since fight or flight behaviors are becoming less necessary in complex culture, the biological mechanisms which produce them are perceived by some power vectors as having outlived their use). The flesh machine must accomplish this task at a rate that at least parallels other pancapitalist demands on the body, demands that cannot be solved through technological intervention. The ebb and flow of drugs used to bring the body up to standard will depend on such developments. The further the flesh machine is behind pancapitalist demands, the greater the use of such drugs (given the mass deployment of stress relievers, it is obvious that the flesh machine has a lot of work left to do). If the flesh machine meets its goal or surpasses it, use of drugs such as these will decrease or even disappear.

Smart Drugs

This class of drugs often complements the use of antidepressants and stress relievers. In the best-case scenario, they consist of nutritional supplements and "natural" medicines which are supposed to enhance memory, stamina, immune systems, and general health. These drugs tend to benefit both the individual and the social system. They pick up where drugs such as stress relievers stop. Rather than bringing the body back to normality, they push the body into supernormality. For consumers, the downside of this category of smart drugs is that their high level of social acceptance creates profitability, which causes a tremendous array of questionable products to appear on the market. Some of these products seem to offer nothing more than a placebo effect (such as the claim that melatonin can reset the biological clock or that ginseng improves sexual performance), while others encourage very profitable useless excess (such as megadoses of vitamins that the body is unable to process). Such corporate cons are bound to parallel any product trend, but what is most interesting is that the desire for smart drugs, which has been manufactured through spectacle to further develop this market, has the potential of expanding their domain, increasing the intersection of smart drugs with other varieties of drugs. In terms of drugs used for body enhancement, those which are developed for medical purposes (to restore normality) and for recreational purposes can be used for "smart" purposes.

When smart drugs are used in this manner, they indicate a victory for pancapitalist socialization, since they are not being used to regulate the maladapted body, and are instead being used to give an individual an "edge" within the space of production/service, indicating an extreme conformity to imperatives of production. The list of possibilities extends from everyday life chemical intervention, like excessive coffee consumption, to illicit and at times counterproductive drugs such as methedrine and cocaine. When used in proper doses, these drugs allow the individual to work longer and harder due to artificial stimulation. Unfortunately, prolonged use of high-powered drugs in everyday life leads to a point of diminishing returns, and finally to counterproduction. This potential problem, combined with a lack of authoritative supervision in the consumption of these substances, places them in a unacceptable category from the perspective of pancapitalist power vectors. However, by submitting to and paying for proper supervision, alternative high-powered drugs can be obtained. The most notable example of the use of social control drugs for body enhancement is the emergence of the "Prozac advantage" among professional and managerial classes. Prozac is among the most commonly prescribed antidepressants. For example, in 1993 in the US, 650,000 prescriptions per month were written for patients. (It is difficult to say how many were actually in need of the drug as a means to cope with physically generated depression or to cope with pathological social conditions, as opposed to those using it for its smart qualities. It should also be noted that any skewing of this statistic toward the middle class is primarily indicative of inequitable distribution of health care.) Given the extremely wide availability of this drug, there are, at the very least, anecdotal cases of the adapted body searching for regulated flesh enhancers that catapult the user into artificially accelerated instrumental action.

Smart drugs aside, the drastic deployment of social control drugs in general indicates a growing intensity in the pathology of social space, from the domestic to the productive. At the same time, it must be noted that a trend toward

social acceptance of the psychological/psychiatric industry as a necessary part of postindustrial life has paralleled
this deployment. Indeed, the growth and empowerment of
medical authority and its industries is currently a structural
social necessity, somewhat akin to the structural need in
early capital to construct a massive bureaucratic class out
of the displaced agrarian class. As late capital has matured
into pancapitalism, all possible measures must be taken to
resituate "significant" people in constantly and rapidly
changing conditions as quickly as possible. Smart drugs
will obviously help in this endeavor. Whenever cyborg
technology fails to meet social demands for body enhancement, biochemical intervention will function as a key
supplement—particularly in maintaining and enhancing
interior ideological inscription.

Steroids

In addition to drugs for mental/behavioral social control are
those which enhance the presentation of the body in
everyday life. Steroids are an unusual example, because
their use went through a significant transition during the
70s and 80s. Steroids were originally and primarily task-
oriented drugs which were used by athletes training to
reach peak bodily performance. For sports which required
inordinate amounts of bulk, strength, and stamina, steroids were a means to accelerate and enhance the training
process. Unfortunately, they rapidly showed themselves to
have side effects that did not necessarily aid in accomplish-
ing the goal at hand. The primary problem was that
steroids produced psychological and behavioral effects
(such as uncontrollable aggression) that were less than

"sportspersonlike." As the use of steroids progressed, the drug began to show serious signs of diminishing returns for the physical body in the form of organ decay. Even without long-term studies on the detrimental effects of steroids, the short-term evidence of physical catastrophe was so overwhelming that steps were taken to eliminate steroid use among those participating in spectacular sports, from the schoolyard to the stadiums. Spectacle had turned against itself, as sports fans watched their heroes die at an early age in exchange for a successful but brief sports career. The situation was rapidly corrected by introducing mandatory drug testing for athletes in institutionally sponsored sports.

Steroids thus became black market drugs. While they still were used as task-oriented body enhancers, this function fell back into a secondary position. The former secondary characteristic rose to the primary position—it was now a drug of body spectacle. Since those who were using the drugs could not compete in organized sports, steroids became a drug for beach-side body builders, and those who desired a sharply cut body. The spectacle of health and vitality (and to some, sexuality) signified by rippling muscles, washboard stomach, and a fat-free body is too deeply etched in culture in the US to eliminate steroid use among elements of the athletic star wannabes, and those who simply wanted to look "perfect." The last holdout of the seemingly task-oriented user is in professional wrestling. Ironically, in professional wrestling, there is no task to accomplish, as the matches are predetermined; however, if an individual desires a career in this theater of flesh, the bodily spectacle must fit perfectly with the designer conception of how a wrestler should look. Task and spectacle implode in a lovely Hollywood moment.

Sacrifices via steroids will be short-lived in the wake of the flesh machine. Disappearing along with the drama of biochemical human sacrifice for the sake of spectacle will be steroids themselves. As flesh products continue to expand, and genetic engineering services become more precise, designing a organic mirror of the spectacularized body, or designing a body predisposed for a certain task will be as easy as taking steroids, although the designer body will be chosen for the individual rather than by the individual.

Weight Loss Drugs

Like steroids, drugs used to induce weight loss have taken radical turns in their development. In the 50s and 60s it was not unusual for individuals to go to physicians and request chemical intervention for the purpose of weight reduction. (Women, of course, were the primary candidates for such treatment). During this period amphetamines were often prescribed. Needless to say, they worked very well to satisfy the desire for weight loss, and even helped to boost production in a variety of social spaces. Unfortunately, the consequences of prolonged use were so negative that this type of medical intervention ceased to be a common practice. Medical intervention for weight reduction was then redirected toward cases of extreme obesity, and other nonpharmaceutical strategies of intervention were pursued. However, the legacy of this period continued in the form of a booming over-the-counter weight loss industry.

By the 70s all the necessary ideological factors were in place for a successful diet industry. Obesity was conclusively linked to a variety of physical pathologies, so if one wanted and valued a longer, healthier life, excessive weight had to be eliminated. In conjunction with this medical imperative came new aestheticized notions about the body. The body beautiful was linked to the absence of fat. Media spectacle relentlessly presented the normalized, attractive body as slim and agile, to the point that the rate of pathologies (anorexia and bulimia) associated with weight loss dramatically increased. In conjunction, the body beautiful was presented in an environment of material abundance, thus indicating a mythic correlation between a sleek figure and wealth. The legacy of the amphetamine diet clicked perfectly with this situation. Consumers wanted an easy method of intervention for weight loss like the ones that had existed in the past. Unfortunately they could no longer turn to medicine, which in turn left a substantial void in this market of desire. Corporations ran to fill the void with a variety of weight loss supplements, thus creating a multimillion dollar industry. The problem was that these supplements were just that, only supplements—a modest addition to a strict diet. Controlled caloric intake and exercise are the only certain means of losing weight, and that method really needs no supplement. In addition, obesity correlates very strongly with genetic predisposition, and one can be sure that no pill will change that. The diet industry, with its array of useless products, is among the greatest market cons perpetrated on the public in this century.

While the profits from body presentation drugs are a valuable contribution to the economy of excess, the ideological contribution of its spectacle is priceless. The diet industry and its allies (the cosmetic and hygiene industries) have managed to redefine bodily beauty as a series of

culturally constructed ideal types. Through exposure to
the spectacular deployment of these representations, a
large segment of the population has been convinced that
the ideal types are replications of material reality, and
hence represent normalized physical bodies. In order to
live up to these impossible standards of aestheticized
normality, consumers must purchase goods and services to
supplement their defective, underaestheticized bodies. Body
presentation supplements are provided to meet any bud-
get, so everyone with an income can participate in the
process of bringing the body up to code. In this case, the
territory must conform to the map. In turn, this imperative
will act as the foundation for both current and future goods
and services provided by the flesh machine.

Pain Control

Drugs for pain control are among the most common medications
used in everyday life. The market for over-the-counter
pain relievers is of staggering size, and the market logic for
drugs which arrest modest pain fits perfectly with the
imperative of body normalization; however, the logic of
drugs used for extreme pain control is very fuzzy. Certainly
this category of drugs (narcotics and analgesics) resists the
process of rationalization in that it cannot be easily cor-
ralled within the limits of instrumentalism. Typically, the
user is provided with pain relievers under the careful
supervision of medical authority. Pain killers are used to
normalize the body by eliminating intense pain so injured
persons can rest during a post-trauma healing process, so
they may be somewhat comfortable during a terminal
illness, or so they can avoid physical debilitation and

maintain normal productivity. When used in these contexts, pain killers seem to benefit both the user and the
social system. They work for the individual by minimizing
an unpleasant and often totalizing factor of experience,
and at times to aid the healing process. They work on
behalf of the social system in the best-case scenario by
allowing an individual to maintain social functionality. In
the worst-case scenario, they work by stopping an excessive flow of empathy from those who are intimate with the
sufferer, and by stopping behaviors that may disturb those
who share a common physical space with the user. For
example, if a patient is convalescing in a hospital and is in
great pain, no one (visitors, medical staff, roommates, etc.)
wants to hear this person screaming in agony. Such behavior is debilitating in every sense for all who are within
hearing, and hence steps will be taken to neutralize the
activity.

The problem with pain killers is that they do not just
neutralize pain—they also produce pleasure and euphoria
in the user, and this is what makes them socially trouble-
some. Pleasure negates some of the socially valuable qualities
of pain killers. For those who must use pain killers to
maintain productivity, the contentment brought about by
the drug is likely to negate productivity by removing an
individual's motivation to work. Also, pain killers of this
class set a very dangerous precedent because they do more
than normalize the body: They give the user the added
bonus of temporary contentment. The worry here is not so
much the Christian fear of pleasure and belief in the
cleansing qualities of pain, but fear of what might occur if
a consumer actually got a taste of satisfied desire (addiction
is one common consequence, objectionable because it

removes the user from market-driven diversified consumption). The problem is doubled when power vectors
realize that users obtained pleasure that was neither
intended nor paid for. The medical response to these
problems has been to relieve the pain only enough to
make it barely tolerable. This policy clearly indicates
that medical intervention for pain favors the demands of
the social system far more than it favors those of the
individual. (It should also be noted that this situation
heavily intersects imperatives framing the "war on drugs").

In the US, the resistance to aggressive pain management, an unthinkable result of the need to perpetuate
pancapitalist ideological imperatives, has reached such
sad proportions that individuals commit suicide rather
than face the medical withholding of pain treatment.
Part of the reason for the current legal debate over an
individual's right to die stems from this very problem.
Not even the flesh machine can imagine how to solve
this conundrum. It can supplement the medical policy
of minimizing the arrest of pain by attempting to develop individuals with higher tolerance for pain, but it
cannot make the pain/body problem go away. Unlike its
control of stress mechanisms, it cannot eliminate or
even reduce pain sensors in the body, since they are
necessary for an individual's survival, and because the
physical ability to feel is a sense around which capital
has produced a massive variety of products. Most unthinkable of all would be to allow the individual to
control he/r own endorphin supply; this would lead to
certain and extreme counterproduction. In terms of
social control, a self-regulating endorphin mechanism
designed by the flesh machine is a possible option, but

the consequences would be the elimination of the fiscal structure of pain products and services already in place. The pharmaceutical industry would fight to stop implementation of such a policy, and most capitalist agencies would realize that the financial loss to the general economy would not be worth the social control this would provide.

The fate of drugs used for bodily/behavioral normalization is uncertain as the flesh machine continues to mature. It seems that such use of drugs will diminish in significance as power vectors become more adept at designing bodies that are predisposed to normative behavior. In matters of normative social activity, prevention of deviance is always a superior strategy to arrest and containment of deviant persons. However, even prevention has its imperfections, and culture itself offers many paths of deviation that cannot be controlled by biological means. To complicate matters further, social conditions are changing at such a fast rate that neither humans nor flesh technologies can keep up with environmental demands. Since lag time for bodily adjustment to new conditions is economically unacceptable under the metaprinciple of efficiency, other means must be found to rapidly bring the body up to code. For these reasons, biochemical intervention by medical authority will remain a significant control strategy, and one can expect that the flesh machine will maintain a research wing dedicated to improving pharmaceuticals designed for social control.

Notes

[1] The primary exception to this scenario is mania, as long as it is directed toward accepted social goals. The excessively energized subject, when focused on production, can often produce higher quality products than his normalized peers, and hence is not a candidate for biochemical intervention. However, the minute the mania becomes nonproductive, or manifests itself as consumption in a manner beyond the subject's financial status, intervention is almost assured.

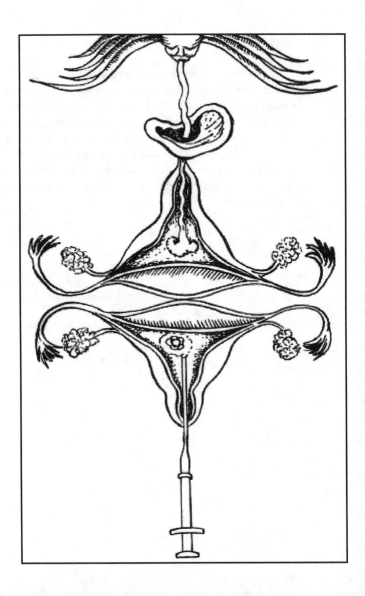

5

As Above, [So Below]*

Faith Wilding

Confession

I desired to have full fruition of my Beloved, and to understand and taste him to the full. I desired that his Humanity should to the fullest extent be one in fruition with my humanity, and that mine then should hold its stand and be strong enough to enter into perfection until I content him, who is perfection itself....To that end I wished he might content me interiorly with his Godhead, in one spirit, without withholding anything from me.... For that is the

[As Above], So Below

Critical Art Ensemble

Confession

More than anything in this world, I wanted to have a child. My gynecologist had always told me that I would never be able to conceive a child by natural means due to blocked fallopian tubes. She suggested adoption to me, but also suggested that if I was prepared financially, and psychologi-

---

*This article was originally published in *Left Curve*, No. 21.

most perfect satisfaction: to grow up in order to be God with God.... In this sense I desired that God give himself to me, so that I might content him....Then it was to me as if we were one without difference. So can the Beloved, with the loved one, each wholly receive the other in all full satisfaction of the sight, the hearing, and the passing away of the one in the other. After that I remained in a passing away in my Beloved, so that I wholly melted away in him and nothing any longer remained to me of myself; and I was changed and taken up in the spirit.

(Hadewijch of Brabant, "Visions," c. l200s)

cally prepared for potential disappointment, I could possibly have my own child given proper medical assistance. I was prepared to do anything, and I did. The process was grueling both psychologically and physically, and the worst part was the harvesting of my eggs and their subsequent implantation. These procedures were as invasive as they were uncomfortable—all variety of surgical instruments cutting, puncturing, sucking, and sliding around my pelvic and vaginal regions. To stay sane, I just kept repeating to myself, "You are going to have a baby." At the end of the process, I cannot describe the excitement, pleasure, and relief when my doctor appeared before me in a glowing white lab coat and said, "You're pregnant." Life was inside of me.

(Anonymous, c. 1990s)

The Paradox of Creation: Temptation and Salvation

Medieval body maps reflect Christian beliefs about the human
    body as a microcosm of the macrocosm—the attunement
    of each organ to a heavenly body, of each bodily fluid to an
    earthly element. Medieval medical practice was largely
    based on the idea of homeopathy, that like cures like, and
    of correspondences: As above, so below. Though much
    homeopathic practice was based on ancient (pagan) sources
    of herbal knowledge, it was adapted to a Christian belief
    system that placed the (souled) human being at the center
    of God's great world plan, and at the center of the cosmos.

Medieval theology had to grapple with paradoxical con-
    ceptions of sexuality, reproduction, and the relationship
    between body and soul. On one hand, the body was

The Paradox of Reproduction

Capitalism has always had an ambivalent attitude toward the
    process of reproduction. On one hand, the economic
    system requires that labor and market populations be
    consistently replenished. On the other hand, the sexual
    activity associated with reproduction has been viewed as
    an unfortunate evil that can detract from the overall
    efficiency of the system—people engaged in sexual behav-
    iors are neither producing nor consuming; rather, they are
    exercising personal sovereignty which, ipso facto, is
    counter-productive and confounds top-down hierarchies.
    This situation has led to a peculiar opposition in which the
    product is embraced but the process is rejected. Unfortu-
    nately, one requires the other, and the problem is doubled

regarded as a necessary evil which housed the soul during its sojourn on earth. In this view, the lapsed body inherited from Adam's fall is a site of temptation and sin; flesh is imperfect and decaying, and must be constantly monitored, controlled, and punished into submission and obedience. On the other hand, the human body was regarded as the pinnacle of God's creation; the body was a means of access to the divine and a means of salvation. As such, it exhibited the Creator's gifts to Adam and Eve—beauty, the senses, and the marvelous powers of generation. These paradoxical readings of the body had to be continually negotiated by the Church which sought to control both the believer's body and soul, and to control human sexual relations through the sacrament of reproductive, heterosexual marriage.

---

because engaging the process does not necessarily yield the product. In turn, various secular attempts have been made by power vectors to streamline sexuality in order to limit it to activities which have some benefit to the political economy.

The primary strategy used by institutions of authority to eliminate sexuality beyond that needed for purposes of reproduction is to label all other sexual practices as deviant, and thereby punishable socially and/or legally. With one mighty blow, gay, lesbian, and all varieties of fetishist sexualities are eliminated from "public" acceptability. While such measures in no way stop individuals forcibly placed in these categories from secretly or defiantly exercising their individual sovereignty, they serve as a reminder

The contradictory medieval conceptions of the body became particularly charged when they were applied to theories and images of sexuality and reproduction. Caroline Walker Bynum has pointed out that medieval body images exhibit a preoccupation with fertility and decay. The Church fathers needed to naturalize the idea of sexuality for reproductive purposes only, and to reinforce motherhood as a redemptive state for women. The figure of Mary was constructed to support this ideology. Modeled partly after desirable characteristics of local pagan goddesses, and partly constructed to accord with programs of religious body control, the image of Mary represents paradoxical and miraculous qualities. She is a virgin, yet fertile; mother of a divine son with whom she is also joined in mystic marriage; and she is an intercessor between heaven and earth. Mary was associated with mother, vessel, and

---

that participation in any activity not compliant with capitalist imperatives will bring punishment(s) from which there is no escape.

To intensify the situation, even straight heterosexuality is in a continuous process of streamlining. The methods employed by the capitalist power vectors against individuals vary in accordance with the person's class position. These methods are most visible in the US—the avant-garde culture of pancapitalist authoritarianism. For the underclass, punishment is aimed almost exclusively at women. The domestic labor required to produce a work force socially engineered to maintain a population intended for low-end service work and/or as a reserve labor army is not considered labor that should be

willing womb—characteristics and qualities that all women of the time were expected to emulate. According to Catholic dogma, as a reward for her virtue, Mary was never subject to decay and she was physically taken into heaven. Thus Mary was reconfigured to fit the Church's strategy to control the female reproductive process, while preserving the idea of the body as the site of purity and salvation.

The figure of Mary was always contrasted with the figure of Eve, who represented the body as a site of temptation and sexual pleasure. God punished Eve's sexual pleasure by afflicting women with painful childbirth, and by subjugating women to their husband's will in all matters. A woman could redeem her flesh only by becoming a mother in Church-sanctioned heterosexual marriage, or through celibate asceticism, in which the body was renounced with the

---

rewarded; rather, population production is punished in part due to its association with sexuality. In spite of the fact that having sex can yield a functional product, underclass women in the US are now increasingly being denied government subsidies for the necessary population production they contribute to the economy. Sexual pleasure is covertly taxed, and underclass women pay the tax by giving their domestic labor to the state free of charge. The current welfare reform acts intensify the situation by doubling the labor demands on underclass women. Not only must they pay their sexuality tax, but typically, they must also work in the service economy at jobs that pay wages below what is necessary to maintain the domestic labor space in which they are enslaved.

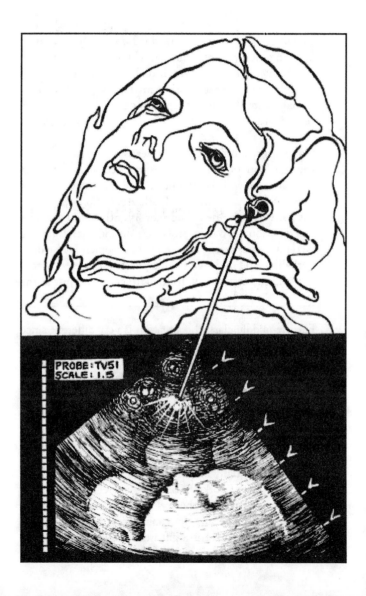

exception of its use to perform good works. The reward for bringing the flesh into spiritual submission was salvation in a heavenly life hereafter. It took rigorous institutional discipline and the creation of a mythology of self-sacrifice to naturalize the idea of separating sexual pleasure from reproduction.

Compulsory motherhood as a means of salvation, and as the only sanctioned way to experience sexuality, went hand in hand with a profound change in the development of medical practice in the Middle Ages. From the 5th to the 13th century, the Church consistently scorned secular intervention in bodily processes; during that time the peasant classes were often treated by women lay healers, herbalists, and midwives, while priests ministered to royalty and the aristocracy. In the 13th century—just as the

---

For the middle class, the situation is very different since wages are generally high enough to subsidize household maintenance. Middle class individuals in the US (whatever their sexual preferences) are threatened by civil law. For example, sexual harassment initiatives in the work place are a tremendous aid to capitalist institutions in eliminating disruptive sexual expressions in the space of production. Since any unwanted sexual expression could be grounds for a civil suit that could cost the perpetrator he/r job and potentially all the wealth s/he has accumulated over he/r years of work, the only survival technique open to individuals is to repress themselves and behave as asexually as possible. To be sure, for capitalist agencies the sexual harassment initiative is a gift from heaven that helps to insure that all employees will engage only in rational and

cult of the Virgin Mary was reaching its zenith—university-trained male doctors began to turn medicine into a quasi-scientific profession from which women were completely banned, causing the loss of their extensive practical experience with and knowledge of the body. Though women still continued to act as midwives and herbalists for centuries thereafter, they were often condemned as witches and put to death for doing so. Thus dual controls, religious and secular, were put in place to ensure that like Mary, women would remain passive and compliant in their relationship to the body.

Even so, compulsory marriage and motherhood seemed a less desirable choice for a minority of women. This was especially true of literate women from the upper echelon of society who entered celibate female communities such as

---

instrumental activity throughout the working day. This situation is doubled with the emergence of victim-driven harassment policies. Here, any sexual behavior an individual witnesses that could be construed as "offensive" must be reported to harassment investigators (literally, bureaucratic sex police) on the premises. Failure to report what could be construed as an act of harassment leaves one potentially liable in the event of a civil lawsuit. In this case, one does not have to be the "victim" or the "perpetrator," s/he only has to witness a sexual expression to be involved in the legal process. This way all employees in institutions with a victim-driven policy are coerced into becoming sex police.

Where then is sexual expression acceptable? It is alive and well in the spectacle. An individual can watch all the

convents or Beguinages. These communities were open to all, so on rare occasions, women of the peasant classes were also accepted into the communities. Some of the women, who practiced extreme voluntary asceticism, holy fasting, and mortification of the flesh as a means to resist compulsory motherhood, presented the Church fathers with a dilemma. On the one hand, in acquiescence to Church dogma, holy women renounced secular human sexual experience as the Church required of those who desired to save themselves from the sins of Eve. On the other hand, these women escaped the instrumentality of reproduction and used their bodies as a means to individual sovereignty and social power. In a truly homeopathic reversal, the body was reformatted as a site for autonomy. The flesh was explored as a means to freedom through sensual presence—female mystics physically embraced God in the

---

Hollywood passion s/he wants, or s/he can have all the cybersex s/he desires. As long as sex is out of the material world, and safely on the screen where it becomes an object of consumption or an object to motivate consumption, it generally stays within the bounds of public acceptability. Sex must not be an act of direct participation; it can only be passively witnessed during leisure hours, if an individual wishes to escape punishment. Hence, individuals of the middle class are caught between spectacular sexuality or state-sanctioned monogamous heterosexuality. By accepting the latter option, individuals are rewarded with relative tolerance of their private, useless sexuality. For the underclass, the situation is worse, as members of this class are limited to spectacular sexuality, because engaging heterosexuality only serves to

ensouled flesh of His decaying creations by tasting the
wounds of lepers and the vomit of the sick, and in feeling
the pain of their own emaciated bodies. Decaying flesh was
transubstantiated in the holy fire of the mystic's desire to
independently commune with God. These powerful (albeit
rare) acts of spiritual rebellion were a theological knot that
the Church patriarchy was at a loss to untie.

Flesh Redeemed: Separating Sin and Creation

In separating sexual pleasure (sin) from reproductive creation,
the relationship of matter and spirit (body and soul) had to
be articulated, and a means of mediating the two orders
had to be created. In order to redeem sinful flesh, Christ
had to become flesh in a redemptive act of creation—a

---

increase the probability of enslavement to the forces
and spaces of production.

Flesh Redeemed:
Separating Sexuality and Reproduction

While pancapitalism's Orwellian anti-sex campaign is certainly
a success, it can always be improved. Improvement is partly
measured by the degree to which sexuality and reproduction are separated. Once separation becomes a legitimized
and accepted element of everyday life, totalized intoler-
ance of sexuality can be initiated in the middle class. The
first experiments in the practical separation of sexuality
and reproduction are currently underway. (Sexuality and

homeopathic strategy. In a structure of correspondence, he became the new Adam and perfected human matter through his birth, death, and resurrection. Mary, as the reprogrammed Eve—the pure vessel, fruitful though not tainted by human fertilization—had the special task of redeeming female bodies, especially the organs of sexual reproduction (*materia mater*). Christ was a virtually conceived embryo that became both human and immortal (resurrected) flesh. Mary was the ethereal flesh machine (the hardware), who interfaced with God (the programmer) through the disembodied Word transmitted by the bodiless angel Gabriel (software). In terms of a reproductive narrative, this is an example of the creation of perfect flesh produced with perfect efficiency: no wasted sperm, no ovulation problems, no failed implantations or blocked fallopian tubes, and no repeated attempts at conception. As noted

---

reproduction have long since been separated symbolically by the division between psychology and biology). By obtaining volunteers for this flesh experiment from pools of individuals intent on having children of their own, but who are unable to do so without medical intervention, medical science hopes to demonstrate that a "better baby" (one better adapted to the imperatives of pancapitalism) can be produced through rationalized intervention. Once such a demonstration occurs, there are empirical grounds for the argument that medical mediation of the total process of reproduction is both desirable and necessary. The promise of a "fitter" child can act as a spectacular resource to convince those members of the middle class not in need of medical intervention to reproduce that separating sexuality from reproduction is beneficial to

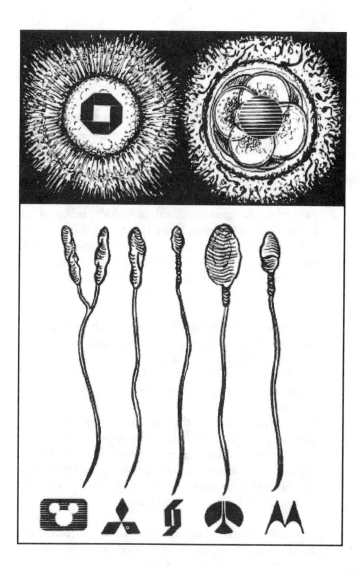

above, after incorporating and giving birth to divinity, Mary's body too became metaflesh which did not die or decay as sinful flesh does, and was taken into heaven for eternity.

The Church understood the need for providing inspirational and concrete representations of the mysteries of divine creation. The narrative of Mary's miraculous conception and virgin birth was encoded in increasingly hyperbolic and beautiful images which served as exemplary and devotional guides for an illiterate lay population. In particular, the great medieval cathedrals dedicated to Mary—such as Chartres, Notre Dame, and Autun—provided ecstatic sensual environments in which soaring architecture, glowing rose windows, colored frescoes, ornate shrines bedecked with jewels and gold leaf, and sublime

---

both parents and offspring. Rather than letting nature take its course in reproduction, representatives of medical science are inserted as mediating efficiency experts. Hence, not only are sexuality and reproduction practically separated, but so are the parents. This way, reproduction better conforms to the capitalist necessity of efficiency: No useless activity occurs in the reproductive process, and less genetic material is wasted. Excess genetic material is reconfigured into a substance for commodified process, as opposed to becoming one of nonrational potential. In this manner, the reproductive process becomes practically reclassified as a purely medical process.

Since the market for rationalized reproduction had already been structurally established before the neces-

music exemplified the rewards of obedience, self-abnega-
tion, and self-surrender.

By following Mary's example and becoming obedient wombs
in sanctioned marriages, women could aspire to transcendence and salvation. In essence, women passively sacrificed
their subjectivity to the church-state. In the ecstatic sur-
render of self to the divine order, the excess of sinful
sexuality was transformed into the excess of instrumental
reproduction. At the same time, the saints of this order
could find solace in the knowledge that they were using
their bodies to further providence, rather than satisfying
their own selfish desires. The belief that they were producing
new Christian souls (soldiers) to populate the earth and carry
out God's plan for creation functioned to help many women
endure painful cycles of endless pregnancy and childbirth.

---

sary methods and technologies became available, the
initial volunteers for the rationalized reproductive experiments serve also to fund further investigation (as
well as providing products for spectacularization). Here,
research actually generates profits, but the essential
element in this experiment is currently not so much
profits as market share. With "nature" functioning as a
prime competitor, seducing people away from construc-
tions of natural reproduction will be difficult. This is
why the product itself and its spectacularization are
currently more important than profits. If the middle
class is not persuaded to accept interventionist practices, the experiment will stagnate, and the desired
practical separation of reproduction and sexuality will
not be fully realized.

The mediation of the Church in this process was an extremely powerful instrument of enforcement, as well as an effective monitoring mechanism. If a woman did not produce a child every year or so, she was answerable to her husband, the priest, and ultimately the whole congregation. Statues and shrines of the Virgin Mary often were covered with the offerings and the messages of women praying for Her intervention in conception and childbirth. In surrendering themselves to the ecstasy of divinely-mediated reproduction, these saints were sustained by the vision of perfect flesh—the resurrected flesh—which was from the beginning the reward promised by Christianity to the devout. In the image of the resurrected flesh, paradoxical aspects of the body are finally transcended and resolved. Like the ecstatic mystic whose body does not decay after death, the resurrected mother's

---

Should the reproductive research wing of the flesh machine fail in its project, the loss to pancapitalism as a whole will be tremendous. The long desired production of a person who uses he/r body solely for purposes of production and consumption (and who is thereby perfectly orderly) will never occur. From the perspective of pancapitalism, nothing less than production of capitalist saints will do. The new Saints of the Pancapitalist Order will be those of perfect flesh. From their genetic code to their cultural code, they will reflect capitalist order and follow its commandments. They will sacrifice their minds and bodies to improve and refine the pancapitalist order. The Saints of the Pancapitalist Order will know a different kind of excess—not one emerging from convivial sociability or erotic, convulsive pleasure, but one dictated by commun-

flesh will be gathered up with the community of Saints to become one perfect body in eternity. And for this ecstasy no sacrifice could be too harsh. Today, this pervasive religious narrative of reproduction as a means of personal salvation and transcendence still lies at the heart of the compulsion for biological reproduction, even though the narrative has become secularized, and the interventions of science have replaced those of divinity.

---

ion with the means of production and by localized proximity to the commodity. The life of a Saint will be one of duty and service to the bureaucratic and the technocratic agencies from which one has received he/r genetic and cultural design. To act against these agencies will be to turn against the Creator—a lost cause suited only for the unfit. All of this the Saints of the Pancapitalist Order will do, and they will do it even if denied a reasonable share in the profits of their production. The reward for their holiness is a higher probability of genetic survival—a promise of life everlasting in which their redeemed flesh conquers the limits of mortality by spreading its canonized code across space and time.

Conclusion

The mythic structure separating sexuality from reproduction/
creation has been fairly constant in the development of
Western culture. The one major disruption is a directional
shift in the ultimate purpose. Currently, the dynamic of
this separation is moving toward the material rather than
the ethereal, toward the rational rather than the
nonrational, and toward the visible rather than the invisible. However, what is truly interesting is not so much the
dynamics of the situation, but the manner in which contingent elements are replaced within the general mythic
structure. The medieval vision of human corruption in
need of intervention has remained. The contingent elements—the institution of intervention and the process by
which successful intervention is obtained—have been
transformed. Rather than the Church, with its connection
to angelic saviors, acting as the institution of redemption
in regard to the sin of sexuality and the finitude of the flesh,
the scientific/medical establishment, with its connection
to nature's Code, has become the institution of mediation
for those who hope to achieve the grace of peaceful
immortality. If maximum access to the secrets and mysteries of the Code is desirable, more is needed than faith in its
omnipresent being. Devotees must also complete the expected round of works required of each individual. Works
are no longer those of rigorous prayer, engaging the sacra-
ments, pilgrimages to sacred sites, self-flagellation, and
asceticism; rather, they have become repetitive work,
power breakfasts, daily commutes (physical or electronic),
fitness training, and sexual self-suppression. The drive
toward immortality through successful reproduction of
perfect offspring requires eternal vigilance and constant

institutional and self-surveillance. While diligently engaging in daily works in no way guarantees access to the Code, it is the only chance for grace. Yet those who are fruitful in their endeavors and collect the necessary assets can buy the desired access to the Code; this in turn, will assure their immortality. In spite of Luther's reformation, indulgences are still the primary currency of salvation.

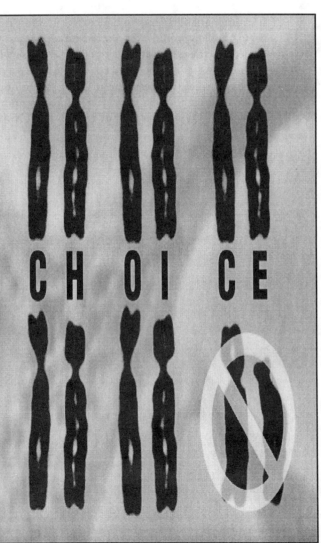

6

Eugenics: The Second Wave*

Eugenics never died after its failed implementation during the early portion of the 20th Century. It has merely been lying dormant until the social conditions for its deployment were more hospitable. Why would it disappear? Eugenics is a perfect complement to the capitalist political-economic imperative of authoritarian control through increased rationalization of culture. Why should the body or the gene pool be sacrosanct? Like a city, a factory, or any other construction of culture, these phenomena can be molded, enhanced, and directed to fit the dominant values of a culture, so that they might efficiently progress into the future. Eugenics, however, is still waiting on the margins of the social, partly because the first wave had a conspirato-

---

*Portions of this article were originally published in *Coil*, No: 4.

rial aura about it. Once eugenics was associated with
Nazi social policy, it was perceived as a top-down manifestation of social intervention and control that reflected
the values of a fascist ruling class, and which negated
democratic principles of choice. Eugenics is also still
waiting in the wings because medical science did not
have the methods and technology to efficiently implement eugenic policy during its first wave (eugenic policy
could only be carried out by mandatory sterilization,
selective breeding, and genocide). Not until medical
science began to radically improve its interventionist
practices (particularly on the microlevel) after World
War II did all the various sectors of culture face a crisis
concerning the limits of organic intervention. While
the public could accept intervention in the process of
dying, intervention in the process of birth was suspect.
To inscribe the body as a machinic system that could be
repaired or maintained through medical and scientific
tinkering was (and is) perfectly fine, as long as medical
science does not attempt to appropriate the role of
creator. For example, to biologically support the immune system through vaccinations that strengthen the
organic system can only be perceived as desirable and
well worth voluntarily acquiring in a secular society,
while creating a new and improved immune system
through genetic intervention is not so desirable (at least
not yet). The goals for eugenicists thus became finding a
way to import the spirit of voluntarism associated with
interventions designed to maintain life into those used to
create it; and, discovering how to construct the perception
that the body, as a machinic system that can be repaired,
maintained, and purified through medical intervention,
can also be improved through genetic intervention.

The eugenic visionary Frederick Osborn already had the answer to these questions as early as the 1930s when he was the director of the Carnegie Institute. Osborn argued that the public would never accept eugenics under militarized directives; rather, time must be allowed for eugenic consciousness to develop in the population. The population would have to come to eugenics rather than vice versa. Further, eugenic consciousness did not have to be aggressively and intentionally micro-manufactured; instead, it would develop as an emergent property as capitalist economy increased in complexity. All that was needed was to simply wait until a specific set of social structures developed to a point of dominance within capitalist culture. Once these structures matured, people would act eugenically without a second thought. Eugenic activity, instead of being an immediately identifiable, monstrous activity, would become one of the invisible taken-for-granted activities of everyday life (much like getting a vaccination).

The set of social structures that Osborn believed had to become dominant were consumer economy and what is now known as the nuclear family. To be sure, both of these social tendencies have come to pass, and are providing the foundation for a more clandestine second wave of eugenic practice. Consumer economy is a necessary foundational component for two reasons. First, if the question of production is solved, and needed goods (water, food, shelter) are generally taken for granted, citizens of the economy of surplus accept all remaining legitimized goods and services as mere purchasable commodities to be chosen or refused. Health care is just another service to be acquired. It becomes neither an unexpected luxury, nor a human right, but just another business component of the economy. Regular medical intervention in everyday life becomes a desirable taken-for-granted service. If eugenic practices are offered as just another commodity under the legitimized authority of medical institutions, as Osborn predicted they would, they too will be taken for granted.

The second foundational characteristic that consumer economy offers is purchase strategies that are based on desire. Consumer economy provides an unending stream of goods, such that a consumer can always desire more. While the wealthiest class can take full advantage of the surplus, and wander into territories of profound waste, uselessness, and excess, the middle class is also offered limited participation. Participation in the rituals of surplus becomes a status symbol, a marker of prestige, a goal-laden value, if not the reason for existence itself. When this economic situation develops in tandem with the rise of the nuclear family, the perception of reproduction begins to significantly change.

It is very clear that the extreme reduction of the family unit is a necessary development in late capitalist economy. The extended family, which functions so well in agrarian-based economies, becomes an anachronism in an economy with a capacity for industrial farming. The situation becomes worse when the extended family is placed in the context of national/global economy; then it actually stops functioning efficiently from the perspective of power vectors, and becomes a detriment to corporate goals. Allowing the extended family to continue offers individuals participating in that institution a social and economic power base which gives them the opportunity to refuse corporate culture. In addition, it creates a social process that has the potential to be more satisfying than participation in consumption processes. Individual loyalty to an institution (i.e., the extended family) that potentially contradicts or negates capitalist imperatives of production and consumption is simply not a possibility that can be allowed to continue. In an effort to eliminate this social possibility, capitalist economy has configured itself to make entrance to or maintenance of middle-class status dependent upon accepting the nuclear family as the model of choice. People are financially rewarded for showing an allegiance to participation in the production and consumption processes, over and above participation in extended family processes.

The process of socializing individuals into nuclear units begins with the education process. Children are immediately taught that "success" in life depends on a division of labor, and on separation from other family members; i.e., the adults work, while the children train in school to enter the workforce. At the end of secondary education, they are fully adjusted to the idea that it is time to leave home to join the workforce, or to attend university. In the US, this process of separation begins almost immediately, because over the past 30 years, production rates have increasingly intensified, while real wages have decreased, thus requiring both parents to work if they want to maintain middle-class status. Children are placed in daycare until it is time for them to attend school. Hence, domestic togetherness in the middle-class family has nearly ceased, and children spend more time with their socializers— education services and mass media—than with "significant others."

The reward for power vectors in promoting this variety of family structure is twofold: First, since people are generally denied social possibilities outside of rationalized contexts, a profound alienation emerges. The only cures offered by capitalist society for this condition are "satisfaction" through success at work, or through acquisition of consumer goods. Second, the geographic mobility necessary for the efficient deployment of the upper echelons of the workforce is assured. People go where their employers send them without a second thought. Whether individuals are near their family or friends is of secondary importance; maintaining class rank (and more and more, simply to remain employed) is of primary importance.

The nuclear family guarantees both the physical and the ideological replication of the workforce; however, in terms of eugenic development, it offers even more. The nuclear family offers a specific set of concerns that complement voluntary eugenics. Since the middle-class nuclear family is generally small, thereby increasing the chances of total familial erasure, its members express a profound concern for reproduction. The extended family is also just as concerned with familial reproduction; the difference between the two, however, is that while the extended family is content with the quantity reproduced as a safeguard of familial survival, the nuclear family is concerned with the "quality" of reproduction. Quality, in this case, is dictated by capitalist demands. Quality means the extent to which a child will be successful, i.e., will be able to obtain a good job in order to maintain or heighten class rank. What nuclear family parents lose in nonrational association with their child, they gain in rationalized association. They can send the child to good schools. They can provide the child with health care. They can offer the child a safe and secure environment in which to mature. The reason parents want to provide their children with these "advantages" is so the child will give society he/r best economic performance. In this thoroughly rationalized situation, *quality of life is equated with economic performance.* The perception is that the better the child performs economically in later life, the better s/he will be able to satisfy he/rself within the structures of production and consumption, and the greater the probability that s/he will be upwardly mobile.

Once the structural conditions of the economy of desire and the nuclear family are in place, which in turn lead to equating quality of life (perhaps even social survival) with economic performance by parents obsessed with their own genetic and/or cultural replication, the environment is ripe for voluntary eugenics—a situation which Osborn was certain would come to pass. If parents are offered goods and services which will give their few offspring a greater opportunity for success, would they not purchase them? Osborn thought that they would, and he believed that these goods and services would include services which would genetically engineer the child to insure he/r better economic performance. He predicted that parents would want to participate in the design of their children to help them to adapt economically and socially—eugenic participation would be a sign of benevolence. To be sure, once eugenics is perceived as a means to empower the child and the parent, it loses its monstrous overtones, and becomes another part of everyday life medical procedure. Capitalism will achieve its goals of genetic ideological inscription, while at the same time realizing tremendous profits for providing the service.

A Brief Note on Class and Eugenics

Traditionally, eugenic ideology has been deployed in the wealthier classes. Cleansing the gene pool of the lower classes has generally been perceived as unnecessary, since the tasks that the lower classes perform are simplistic and therefore almost any genetic configuration will do. Most likely, traces of this ideological tendency will continue in regard to the working class. At the same time, however, eugenic ideology will be vigorously deployed down the class scale, until a point is reached where the purchase of the services is no longer financially possible. Unlike in the past, power vectors believe including all levels of the middle class in genetic design to be more essential than ever, so that all "significant" populations can make the "evolutionary" jumps necessary to keep abreast of rapid cultural development.

The working class will probably not be called to participate in the new wave of eugenic practice. Since the poor are reproducing at a rate beyond that needed to keep low-end labor conditions stable, no reason exists for power vectors to construct interventions in their replication process (perhaps with the exception of slowing it down). In the US, it is riduculous to think that members of the lower classes—who are not even granted health care—will be able to participate in costly eugenic practices. Currently, infant mortality among the poor is absurdly high simply because of a lack of prenatal care, so it seems unlikely that the lower classes will be presented with less necessary elements of "medical care." In European nations, where health care is provided for all citizens, a different scenario could emerge. Eugenic practices may be promoted all the way down the class scale. Much depends on whether or not eugenics delivers on its promise to rationalize the gene pool in a way that seems economically and socially productive to capitalist forces. Should eugenics fulfill its promises, the US would also have to comply with full-scale deployment, in order to stay competitive in the global economy.

Another element that will affect the deployment of eugenic practices will be the degree to which cyborg technology seeps down into the lower classes. If organic platforms are needed for duties below those filled by members of the middle classes, then eugenic deployment could go all the way down the class scale. However, this scenario seems unlikely, as the past record shows that when modified by technology, working class tasks tend either to go completely robotic or shift to a smaller number of low-end technocrats.

More Utopian Promises

As one would expect, eugenic practices are already receiving mass media support in an effort to build eugenic consciousness in consumers. Certainly, "eugenics," "genetic cleansing," or any other term suggesting the horror of the first wave of eugenics is never mentioned in these moments of spectacle, and the spectacularized narratives of bio-tech are presented to individuals in a seductive rather than a forceful way. For example, a consumer can purchase genetic testing (cleansing) services that promise to assure the parent of a healthier child. At the four-to-eight-cell stages, an embryo can be tested for a variety of genetic diseases and deformations. Some genetic defects can be repaired. At the very least, a defective embryo can be terminated, and the parents can try again to produce a healthy, normalized one. Of course, no one is forced to take the test (it must be desired and purchased), and if any abnormality is found, no one is forced to terminate the creature. One can even choose to let the creature grow to the 16-cell stage, at which time it will self-terminate if it is not implanted in a uterus (perfectly natural). As promised, services such as this one allow concerned (obsessive) parents greater assurance that their child will be normal and healthy, and that they will be spared the financial and psychological burden of an abnormal child. The subtext, however, is just as Osborn predicted: The parents make the decision regarding termination in accordance with the imagined child's probability of success in life. They choose to accept or terminate the imagined child, not so much to fulfill their own needs as to fulfill the needs of pancapitalist culture. In spite of all the can-do spectacle regarding the productive and happy lives of the "differently-abled," the emphasis here is not on the "happy" (the nonrational) but on the "productive" (the rational). To be sure, "healthy" and "normal" correlate with the projected potential of the imagined child's productivity, combined with the parents' continued need to participate in particularized modes of consumption that do not include purchasing goods and services for the defective. Rational patterns of production and consumption in the economy of desire are presented as determinants of a happy parent-child relationship, instead of the happy parent-child relationship being determined by nonrational characteristics such as love, concern, and understanding. If the parent-child relationship were based on these latter qualities, and not those of potential production and consumption, what need would there be for the test in the first place? The spectacle promises its viewers that testing benefits the parents and child by eliminating sickness, but what these half-truths lead to is a eugenic consciousness that serves ideological directives implanted in consciousness by pancapitalist initiatives.

The spectacle of reproductive bio-tech also promises to assure fertility in a majority of cases. Even if a reproductive system is in disrepair, it can be technologically modified and/or coaxed to function as expected. The demand for such technological insurance is peculiar, since there is no shortage of children in need of a parent. Certainly, nonrational beliefs explain much of this economic riddle: Perhaps parents value participation in the "magic" of the reproductive process; perhaps they want to see their own physical characteristics duplicated in the next generation; or perhaps successful reproduction validates their (essentialized) gender positions. The list of entries and the manner in which they can be combined is quite extensive, but not exhaustive. While nonrational associations with reproduction are useful in selling reproductive goods and services, rational concerns also come into play. Would-be parents tend to find it desirable to have total control over the physical care and early socialization of the child, so they can be certain that nothing can disrupt the future success of the child. The only way to have this assurance is to be a primary participant in these processes from conception until the child is turned over to the education system. (This would, in part, explain why obtaining genetic materials from outside sources is preferable to adoption).

One must also ask, why are there problems with individual fertility in the first place? Much of the answer lies outside the realm of cultural design, but part of the answer lies in the economy of investment for medical research: In regard to funding, research which could help to prevent infertility takes second place behind research that can insure fertility. (For example, funding for research aimed toward eliminating pelvic inflammatory disease, which can cause infertility in some women, is relatively meager when compared to investments in research to create products and services for assisted pregnancy). This funding tendency creates an expanded demand for the fertility products and services by underfunding research that could lead to a cure for root causes of infertility. Rather than investing in research that could produce preventive care, funding agencies invest in research to develop more profitable means to repair an injured reproductive system. In turn, the increased likelihood that women will need assisted reproductive care channels the target population into medical institutions where they are likely to engage additional reproductive services.

Extending fertility has similar consequences. This utopian promise does seem desirable for women in many ways. If reproductive assistance can increase the span of years during which a woman can reproduce, she would have far greater choice in how to plan her life. (Currently, the fertility range has not been significantly altered, since the success rate for assisted pregnancy drops dramatically after the age of 40). If a woman knew she was able to have a child after age 40, it would allow her uninterrupted time to establish herself in the workforce and acquire the wealth needed to best provide for the child. The option of being both a successful mother and a professional woman would increase in likelihood. Obviously, the state would also benefit by delaying reproduction to later years (a trend which is occurring among middle-class women), since there is a greater structural demand for women to enter the workforce, and deferral of reproduction would allow them to function better within it. In addition, the prevalence of middle-aged pregnancy would channel (middle-class) women into medical institutions where they would be most likely to engage in voluntary eugenic practices. As with most seeming social benefits, the majority of them are gains for the state, while those the individual receives are primarily incidental consequences of state sanctioned social policy.

The Spectacle of Anxiety

The spectacle of anxiety also hides itself in utopian spectacle, but rather than aiming the presentation at individuals, this spectacle is normally directed at social aggregates. For example, there is considerable coverage of breakthroughs in medical science in media ranging from knowledge-specific journals to popular newscasting. The most glamorous subjects tend to be concerned with the rationalization of death (cancer, heart disease, AIDS, and so on), but genetic research, concerned with the rationalization of birth, also makes the list. For the most part, these discoveries are framed by a national identity. On the individual level, the nationality of the scientists who made a given breakthrough is fairly irrelevant, and most are relieved that medical science is constructing a healthier tomorrow. However, at the national level, who discovered what has very deep economic implications. Each announcement of a surge in applied medical science that is beyond the national borders represents lost profits and an increase in the national research gaps. (The real loss, of course, is to other competing multinationals, rather than to nation states). The public perception of losing national economic advantage is a tremendous fuel to create a popular consensus for high-velocity research (a permanent corporate R&D policy, whether the public agrees or not) as opposed to cautious and critical low-velocity research. As with the individual purchase of goods and services that offer an economic advantage, will the development of goods and services that are perceived to give a nation an economic advantage also be pursued without question? This has certainly been the case in the past, and continues to be true now. Such a situation seems to indicate that the time is right for eugenic practices to flourish on the macro as well as on the micro levels of society.

Jamming the Eugenic Failsafes

In addition to utopian promises, medical science makes numerous ethical promises to the public designed to reassure populations that the eugenic beast will not be reborn. As far as involuntary eugenics is concerned, these promises have merit, although the promise not to engage in state-sanctioned involuntary eugenic practices is an easy one to keep, since the strategies to develop privatized voluntary eugenic practices are proceeding so smoothly. On the other hand, the ethical promises to forbid practices which either lay the foundation for the implementation of voluntary eugenic policy, or which are eugenic in and of

themselves, can be looked upon with a great deal of skepticism. For example, one key promise from medical science is that human organic matter will not and cannot be sold. In some cases, medical science has lived up to this promise. In the case of organ sales, there are other options to pursue, such as artificial, cloned, and transgenic organs (all of which are still in various stages of experimentation). These organ replacement products can be sold. The promise of zero sales of human organs is also fortified by the fact that it is difficult to find donors willing to sell their organs, since doing so will either kill them or decrease their life expectancy. However, with human reproductive matter, the situation is much different. Sperm and eggs can be harvested without threatening the life of the provider. In this situation, medical science has *legally* kept its promise. Sperm, eggs, embryos, etc., are not being bought and sold; they are being donated. However, while the organic matter cannot be bought and sold, the harvesting and the implanting processes are salable services. The medical establishment has jammed this ethical failsafe simply by building the fiscal structure of the industry around the process, rather than around the product.

To make matters worse, eugenic screening practices are used to acquire suitable reproductive materials. Potential donors are thoroughly tested physically and psychologically to make sure they meet industry standards of health and normalcy. Family histories are acquired and scrutinized so that those receiving the materials can be sure that there are no latent genetic defects that could lead to a problematic outcome. If a potential donor is found to be suitably pure, then s/he can become an actual donor. Of course, no clinic would admit that it is constructing a pure gene pool—a purity which is dictated by the political and economic demands of pancapitalism. Rather, such institutions claim that they are only attempting to provide consumers with top value for their purchasing dollar, and preserving their own reputations as institutions of high integrity that provide high-quality products and services. Screening is done for economic purposes, and not for political purposes. To an extent this is true. It seems very unlikely that conspiratorial teams of doctors are plotting a new master race; however, just as Osborn predicted, eugenic mechanisms are emerging out of the rationalized reproductive process which reflect the ideological values of the social context in which the process occurs (the primary value, as Osborn believed would come to pass in consumer economy, is that people's value is determined by their economic potential).

This same process is replicated in the implementation of selective reduction. To increase the probability of a successful implantation procedure, a small set of embryos (three to eight) is placed into the uterus; the number depends on the quality of the embryos and the age of the woman. The results vary; however, the probability of successful implantation (when a embryo attaches itself to the uterine wall) is increased. At times, the procedure is too successful, and produces more than one fetus. This leaves the client with the choice of bringing all the fetuses to term, or of reducing their number. Many times, the reduction is necessary as the number of fetuses conceived could pose a threat to the life of the client, but just as often, fetus reduction is implemented because the client desires a specific number of fetuses. The client can select (often in accordance with viability) which fetuses she wants to keep. In the cases where the fetuses are equally viable, the client can select for aesthetic characteristics (such as the number of children, the gender, or the gender combination). Like donor screening, there is nothing genetically conspiratorial about the process; clients are simply purchasing the specific goods that they want. Yet once again, the desire for a specific product is manufactured by spectacle that is directed by ideological as well as marketing concerns. The process of *selective* womb cleansing is political and eugenic, and is an emergent byproduct of rationalized reproduction.

Conclusion

Osborn's predictions are coming to pass. The time is right for the second wave of eugenics because the economic foundation has been laid. Eugenics complements the grand pancapitalist principle of the total rationalization of culture. The foundation for consumer consciousness is replicated in the foundation for eugenic consciousness. Reproduction is spectacularly represented and publicly perceived as an object of surplus that can be produced to meet consumer desire. Desire itself does not emerge from within, but is imposed from without by the spectacular engines of pancapitalist ideological inscription. However, the situation has yet to reach catastrophic proportions. Eugenic practices are still crude and experimental; they still have to work their way across class levels and down the class ladder. Thus far, power vectors have not been able to turn perception into activity (the product is recognized, but few are buying). In order to truly accomplish the goal of making eugenic activity a part of everyday life, the public must be convinced that rationalized processes of reproduction are superior and more desirable than the nonrational means of reproduction. In other words, large segments of the population (with an emphasis on the middle class) must still be channeled into this frontier market. This will take time, during which counternarratives and resistant strategies and tactics can be developed. Unfortunately, in order to seduce all who look upon it, eugenics has masked itself in the utopian surface of free choice and progress. In this sense, power vectors have stolen and are cautiously using the strategy of subversion in everyday life to create a silent flesh revolution.

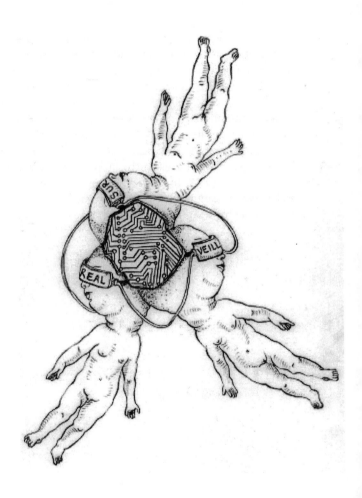

Appendix:
Utopian Promises—Net Realities*

The need for Net criticism certainly is a matter of overwhelming urgency. While a number of critics have approached the new world of computerized communications with a healthy amount of skepticism, their message has been lost in the noise and spectacle of corporate hype—the unstoppable tidal wave of seduction has enveloped so many in its dynamic utopian beauty that little time for careful reflection is left. Indeed, a glimpse of a possibility for a better future may be contained in the new techno-apparatus, and perhaps it is best to acknowledge these possibilities here in the beginning, since Critical Art Ensemble (CAE) has no desire to take the position of the neoluddites who believe

---

*This article was originally an address to Interface 3 in Hamburg 1995, and was published in the conference proceedings.

that the techno-apparatus should be rejected outright, if not destroyed. To be sure, computerized communications offer the possibility for the enhanced storage, retrieval, and exchange of information for those who have access to the necessary hardware, software, and technical skills. In turn, this increases the possibility for greater access to vital information, faster exchange of information, enhanced distribution of information, and cross-cultural artistic and critical collaborations. The potential humanitarian benefits of electronic systems are undeniable; however, CAE questions whether the electronic apparatus is being used for these purposes in the *representative* case, much as we question the political policies which guide the Net's development and accessibility.

This is not the first time that the promise of electronic utopia has been offered. One need only look back at Brecht's critique of radio to find reason for concern when such promises are resurrected. While Brecht recognized radio's potential for distributing information for humanitarian and cultural purposes, he was not surprised to see radio being used for the very opposite. Nor should we be surprised that his calls for a more democratic interactive medium went unheeded.

During the early 1970s, there was a brief euphoric moment during the video revolution when some believed that Brecht's call for an interactive and democratic electronic medium was about to be answered. The development of home video equipment led to a belief that soon everyone who desired to would be able to manufacture their own television. This seemed to be a real possibility. As the cost of video equipment began to drop dramatically, and cable set-ups offered possibilities for distribution, electronic utopia seemed immanent, and yet, the home video studio never came to be. Walls and boundaries confounding this utopian dream seemed to appear out of nowhere. For instance, in the US, standards for broadcast quality required postproduction equipment that no one could access or afford except capital-saturated media companies. Most cable channels remained in the control of corporate media, and the few public access channels fell into the hands of censors who cited "community standards" as their reason for an orderly broadcast system. While production equipment did get distributed as promised, the hopes of the video utopianists were crushed at the distribution level. Corporate goals for establishing a new market for electronic hardware were met, but the means for democratic cultural production never appeared.

Now that giddy euphoria is back again, arising in the wake of the personal computer revolution of the early 80s, and with the completion of a "world-wide" multi-directional distribution network. As to be expected, utopian promises from the corporate spectacle machine saturate the everyday lives of bureaucrats and technocrats around the first world, and once again there seems to be a general belief—at least within technically adept populations—that this time the situation will be different. And to a degree, this situation is different. There is an electronic free zone (the aggregate of domains that have characteristics resistant to pancapitalism), but from CAE's perspective, it is only a modest development at best. By far the most significant use of the electronic apparatus is to keep order, to replicate dominant pancapitalist ideology, and to develop new markets.

At the risk of redundantly stating the obvious, CAE would like to recall the origins of the internet. The internet is war-tech that was designed as an analog to the US highway system (yet another product which stemmed from the mind of the military, and which was primarily intended as a decentralized aid to mobilization). The US military wanted an apparatus that would preserve command structure in the case of nuclear attack. The answer was an electronic network capable of immediately rerouting itself if one or more links were destroyed, thus allowing surviving authorities to remain in communication with each other and to act accordingly. With such an apparatus in place, military authority could be maintained, even through the worst of catastrophes. With such planning at the root of the internet, suspicion about its alleged anti-authoritarian characteristics must occur to anyone who takes the time to reflect on the apparatus. It should also be noted that the decentralized characteristics for which so many praise the Net did not arise out of anarchist intention, but out of nomadic military strategy.

Research scientists were the next group to go online after the military. While it would be nice to believe that their efforts on the Net were benign, one must question why they were given access to the apparatus in the first place. Science has always claimed legitimacy by announcing its "value-free" intentions to search for the truth of the material world; however, this search costs money, and hence a political economy with a direct and powerful impact on science's lofty goals of value-free research enters the equation. Do investors in scientific research offer money with no restrictions attached? This seems quite unlikely. Some type of return on the investment is implicit in any demand from funding institutions. In the US, the typical demand is either theory or technology with military applications or applications that will strengthen economic development. The greater the results promised by science in terms of these two categories, the more generous the funding. In the US, not even scientists get something for nothing.

The need for greater efficiency in research and development opened the new communication systems to academics, and with that development, a necessary degree of disorder was introduced into the apparatus. Elements of free zone information exchange began to appear. But as this system developed, other investors, most notably the corporations, demanded their slice of the electronic pie. All kinds of financial business was conducted on the Net with relatively secure efficiency. As the free zone began to grow, the corporations realized that a new market mechanism was growing with it, and eventually the marketeers were released onto the Net. At this point, a peculiar paradox came into being: Free market capitalism came into conflict with the conservative desire for order. It became apparent that for this new market possibility to reach its full potential, authorities would have to tolerate a degree of chaos. This was necessary to seduce the wealthier classes into using the Net as site of consumption and entertainment, and second, to offer the Net as an alibi for the illusion of social freedom. Although totalizing control of communications was lost, the overall cost of this development to governments and corporations was minimal, and in actuality, the cost was nothing compared to what was gained. Thus was born the most successful repressive apparatus of all time; and yet it was (and still is) successfully represented under the sign of liberation. What is even more frightening is that the corporation's best allies in maintaining the gleaming utopian surface of cyberspace are some of the very populations who should know better. Techno-utopianists have accepted the corporate hype, and are now disseminating it as the reality of the Net. This regrettable alliance between the elite virtual class and new age cybernauts is structured around five key virtual promises. These are the promised social changes that seem as if they will occur at any moment, but never actually come into being.

Promise One: The New Body

Those of us familiar with discourse on cyberspace and virtual reality have heard this promise over and over again, and in fact there is a kernel of truth associated with it. The virtual body is a body of great potential. On this body we can reinscribe ourselves using whatever coding system we desire. We can try on new body configurations. We can experiment with immortality by going places and doing things that would be impossible in the physical world. For the virtual body, nothing is fixed and everything is possible. Indeed, this is the reason why hackers wish to become disembodied consciousnesses flowing freely through cyberspace, willing the idea of their own bodies and environments. As virtual reality improves with new generations of computer technology, perhaps this promise will come to pass in the realm of the multisensual; however, it is currently limited to gender reassignment on chat lines, or game-boy flight simulators.

What did this allegedly liberated body cost? Payment was taken in the form of a loss of individual sovereignty, not just from those who use the Net, but from all people in technologically saturated societies. With the virtual body came its fascist sibling, the data body—a much more highly developed virtual form, and one that exists in complete service to the corporate and police state. The data body is the total collection of files connected to an individual. The data body has always existed in an immature form since the dawn of civilization. Authority has always kept records on its underlings. Indeed, some of the earliest records that Egyptologists have found are tax records. What brought the data body to maturity is the technological apparatus. With its immense storage capacity and its mechanisms for quickly ordering and retrieving information, no detail of social life is too insignificant to record and to scrutinize. From the moment we are born and our birth certificate goes online, until the day we die and our death certificate goes online, the trajectory of our individual lives is recorded in scrupulous detail. Education files, insurance files, tax files, communication files, consumption files, medical files, travel files, criminal files, investment files, files into infinity....

The data body has two primary functions. The first purpose serves the repressive apparatus; the second serves the marketing apparatus. The desire of authoritarian power to make the lives of its subordinates perfectly transparent achieves satisfaction through the data body. Everyone is under permanent surveillance by virtue of their necessary interaction with the marketplace. Just how detailed data body information actually may be is a matter of speculation, but we can be certain that it is more detailed than we would like it to be, or care to think.

The second function of the data body is to give marketeers more accurate demographic information to design and create target populations. Since pancapitalism has long left the problem of production behind, moving from an economy of need to an economy of desire, marketeers have developed better methods to artificially create desires for products that are not needed. The data body gives them insights into consumption patterns, spending power, and "lifestyle choices" of those with surplus income. The data body helps marketeers find you, and provide for your lifestyle. The postmodern slogan, "You don't pick the commodity; the commodity picks you" has more meaning than ever.

But the most frightening thing about the data body is that it is the center of an individual's *social* being. It tells the members of officialdom what our cultural identities and roles are. We are powerless to contradict the data body. Its word is the law. One's organic being is no longer a determining factor, from the point of view of corporate and government bureaucracies. Data have become the center of social culture, and our organic flesh is nothing more than a counterfeit representation of original data.

Promise 2: Convenience

Earlier this century, the great sociologist Max Weber explained why bureaucracies work so well as a means of rationalized social organization in complex society. In comparing bureaucratic practice to his ideal-type, only one flaw appears: Humans provide the labor for these institutions. Unfortunately humans have nonrational characteristics, the most notorious of which is the expression of desire. Rather than working at optimum efficiency, organic units are likely to seek out that which gives them pleasure in ways that are contrary to the instrumental aims of the bureaucracy. All varieties of creative slacking are employed by organic units. These range from work slowdowns to unnecessary chit-chat with one's fellow employees. Throughout this century policy makers and managerial classes have concerned themselves with developing a way to stop such activities in order to maximize and intensify labor output.

The model for labor intensification came with the invention of the robot. So long as the robot is functional, it never strays from its task. Completely replacing humans with robots is not possible, since so far, they are only capable of simple, albeit very precise, mechanical tasks. They are data driven, as opposed to the human capacity for concept recognition. The question then becomes how to make humans more like robots, or to update the discourse, more like cyborgs, thereby getting the best of the mechanical and the organic. At present, much of the technology necessary to accomplish this goal is available, and more is in development. However, having the technology, such as telephone headsets or wearable computers, is not enough. People must be seduced into wanting to wear them, at least until the technology evolves that can be permanently fixed to their bodies.

The means of seduction? Convenience. Life will be so much easier if we only connect to the machine. As usual there is a grain of truth to this idea. I can honestly admit that my life has been made easier since I began using a computer, but only in a certain sense. As a writer, it is easier for me to finish a paper now than it was when I used pen and paper or a typewriter. The problem: Now I am able to (and therefore, must) write two papers in the time it used to take to produce one. The implied promise that I will have more free time because I use a computer is false.

Labor intensification through time management is only the beginning. There is still another problem to be solved in regard to total utility: People can still separate themselves from their work stations—the true home of the modern day cyborg. The seduction continues, persuading us that we should desire to carry our electronic extensions with us all the time. The latest commercials from AT&T are the perfect representation of consumer seduction. They promise: "Have you ever sent a fax....from the beach? You will." or "Have you ever received a phone call....on your wrist watch? You will." This commercial is most amusing. There is an image of a young man who has just finished climbing a mountain, and is watching a sunset. At that moment his wife calls on his wrist phone, and he describes the magnificence of the sunset to her. Now who is kidding who? Is your spouse going to call you while you are mountain climbing? Are you going to need to send a fax while lounging on the beach? The corporate intention for deploying this technology (in addition to profit) is so transparent, it's painful. The only possible rejoinder is: "Have you ever been at a work station....24 hours a day, 365 days a year? You will." Now the virtual sweat shop can go anywhere you do!

Another telling element in this representation is that the men in these commercials are always alone. (This is a gendered element which CAE is sure has not failed to catch the attention of feminists, although CAE is unsure as to whether it will be interpreted as sexism or a stroke of luck). In this sense, the problem is doubled: Not only is the work station always with you, but social interaction will always be fully mediated by technology. This is the perfect solution to abolish that nuisance, the subversive environment of public space.

Promise 3: Community

Currently in the US, there is no more popular buzz word than "community." This word is so empty of meaning that it can be used to describe almost any social manifestation. For the most part, it is used to connote sympathy with or identification with a particular social aggregate. In this sense, one hears of the gay community or the African-American community. There are even oxymorons, such as the international community. Corporate marketeers from IBM to Microsoft have been quick to capitalize on this empty sign as a means to build their commercial campaigns. Recognizing the extreme alienation that afflicts so many under the reign of pancapitalism, they offer Net technology as a cure for a feeling of loss that has no referent. Through chat lines, news groups, and other digital environments, nostalgia for a golden age of sociability that never existed is replaced by a new modern day sense of community.

This promise is nothing but aggravating. There is not even a grain of truth in it. If there is any reason for optimism, it is only to the extent mentioned in the beginning of this lecture; that is, the Net makes possible a broader spectrum of information exchange. However, anyone with even a

basic knowledge of sociology understands that information exchange in no way constitutes a community.
Community is a collective of kinship networks which
share a common geographic territory, a common history,
and a value system—one usually rooted in a common
religion. Typically, communities are rather homogenous,
and tend to exist in the historical context of a simple
division of labor. Most importantly, communities embrace nonrational components of life and of consciousness.
Social action is not carried out by means of contract, but
by understandings, and life is certainly not fully mediated
by technology. In this sense, the connection between
community and Net life is unfathomable. (CAE does not
want to romanticize this social form, since communities
can be as repressive and/or as pathological as any society.)

Use of the Net beyond its one necessary function (i.e.,
information gathering), is, from CAE's perspective, a
highly developed anti-social form of interacting. That
someone would want to stay in his or her home or office
and reject human contact in favor of a textually mediated
communication experience can only be a symptom of
rising alienation, not a cure for it. Why the repressive
apparatus would want this isolation to develop is very
clear: If someone is online, he or she is off the street and out
of the gene pool. In other words, they are well within the
limits of control. Why the marketing apparatus would
desire such a situation is equally clear: The lonelier people
get, the more they will have no choice but to turn to work
and to consumption as a means of seeking pleasure.

In a time when public space is diminishing and being
replaced by fortified institutions such as malls, theme

parks, and other manifestations of forced consumption that pass themselves off as locations for social interaction, shouldn't we be looking for a sense of the social (that is, to the extent still possible), direct and unmediated, rather than seeing these anti-public spaces replicated in an even more lonely electronic form?

Promise 4: Democracy

Another promise eternally repeated in discourse on cyberspace is the idea that the electronic apparatus will be the zenith of utopian democracy. Certainly, the internet does have some democratic characteristics. It provides all its cyber-citizens with the means to contact all other cyber-citizens. On the Net, everyone is equal. The shining emblem of this new democracy is the World Wide Web. People can construct their own home pages, and even more people can access these sites as points of investigation. This is all well and good, but we must ask ourselves if these democratic characteristics actually constitute democracy. A platform for individual voices is not enough (especially in the Web where so many voices are lost in the clutter of data debris). Democracy is dependent on the individual's ability to *act* on the information received. Unfortunately, even with the Net, autonomous action is still as difficult as ever.

The difficulty here is threefold: First, there is the problem of locality and geographic separation. In the case of information gathering, the information is only as useful as the situation and the location of the physical body allows. For example, a gay man who lives in a place where homophobia reigns, or even worse, where homosexual practice is an illegal activity, is still be unable to openly act on his desires, regardless of the information he may gather on the Net. He is still just as closeted in his everyday life practice, and is reduced to passive spectatorship in regard to the object of his desire, so long as he remains in a repressive locality.

The second problem is one of institutional oppression. For example, no one can deny that the Net can function as a wonderful pedagogical tool and can act as a great means for self-education. Unfortunately, it has very little legitimacy in and of itself as an educational institution. The Net must be used in a physical world context under appropriate supervision for it to be awarded legitimacy. In the case of education, in order for the knowledge-value gained from the Net to be socially recognized and accepted, it must be used as a tool within the context of a university or a school. These educational contexts are fortified in a manner to maintain a status-quo distribution of education. Consequently, one can acquire a great deal of knowledge from the Net, but still have no education capital to be exchanged in the marketplace. In both of these cases, there must be a liberated physical environment if the Net is to function as a supplement to democratic activity.

The final problem is that the Net functions as a disciplinary apparatus through the use of transparency. If people feel that they are under surveillance, they are less likely to act in a manner that is beyond normalized activity; that is, they are less likely to express themselves freely, and to otherwise act in a manner that could produce political and social changes within their environments. In this sense, the Net serves the purpose of *negating* activity rather than encouraging it. It channels people toward orderly homogeneous activity, rather than reinforcing the acceptance of difference that democratic societies need.

To be sure, there are times when transparency can be turned against itself. For example, one of the reasons that the PRI party's counteroffensive against the Zapatistas did not end in total slaughter was the resisting party's use of the Net to keep attention focused upon its members and its cause. By disallowing the secret of massacre, many lives were saved, and the resistant movement could continue. Much the same can be said about the stay of execution won for Mumia Abu Jamal. The final point here is that it must be remembered that the internet does not exist in a vacuum. It is intimately related to all kinds of social structures and historical dynamics, and hence its democratic structure cannot be realistically analyzed as if it were a closed system.

Taking a step back from the insider's point of view, achieving democracy through the Net seems even less likely considering the demographics of the situation. There are five and a half billion people in the world. Over a billion barely keep themselves alive from day to day. Most people don't even have telephones, and hence it seems very unlikely that they will get computers, let alone go online. This situation raises the question, is the Net a means to democracy, or simply another way to divide the world into haves and have-nots? We also must ask ourselves, how many people consider the Net really relevant in their everyday lives? While CAE believes that it is safe to assume that the number of Net users will grow, it seems

unlikely that it will grow to include more than those who
have the necessary educational background, and/or those
who are employed by bureaucratic and technocratic agencies.

CAE suggests that this elite stronghold will remain, and
that most of the first world population that will become a
part of the computer revolution will do so primarily as
passive consumers, rather than as active participants.
They will be playing computer games, watching interactive TV, and shopping in virtual malls. The stratified
distribution of education will act as the guardian of the
virtual border between the passive and the active user, and
prevent those populations participating in multidirec-
tional interactivity from increasing in any significant
numbers.

Promise 5: New Consciousness

Of all the Net hype, this promise is perhaps the most insidious,
since it seems to have no corporate sponsor (although
Microsoft has tapped the trend to some extent). The
notion of the new consciousness has emerged out of new
age thinking. There is a belief promoted by cyber-gurus
(Timothy Leary, Jaron Lanier, Roy Ascott, Richard
Kriesche, Mark Pesci) that the Net is the apparatus of a
benign collective consciousness. It is the brain of the
planet which transforms itself into a cosmic mind through
the activities of its users. It can function as a third eye or
sixth sense for those who commune with this global
coming together. This way of thinking is the paramount
form of ethnocentrism and myopic class perception. As
discussed in the last section, the third world and most of

the first world citizenry are thoroughly marginalized in this divine plan. If anything, this theory replicates the imperialism of early capitalism, and recalls notions such as manifest destiny. If new consciousness is indicative of anything, it is the new age of imperialism that will be realized through information control (as opposed to the early capital model of military domination).

Of the first four promises examined here, each has proven on closer inspection to be a replication of authoritarian ideology to justify and put into action greater repression and oppression. New consciousness is no exception. Even if we accept the good intentions and optimistic hopes of the new age cybernauts, how could anyone conclude that an apparatus emerging out of military aggression and corporate predation could possibly function as a new form of terrestrial spiritual development?

Conclusion

As saddened as CAE is to say it, the greater part of the Net is capitalism as usual. It is a site for repressive order, for the financial business of capital, and for excessive consumption. While a small part of the Net may be used for humanistic purposes and to resist authoritarian structure, its overall function is anything but humanistic. In the same way that we would not consider an unregulated bohemian neighborhood to be representative of a city, we must also not assume that our own small free zone domains are representative of the digital empire. Nor can we trust our futures to the empty promises of a seducer that has no love in its heart.

AUTONOMEDIA NEW AUTONOMY SERIES
Jim Fleming & Peter Lamborn Wilson, Editors

T.A.Z.
The Temporary Autonomous Zone
Hakim Bey

THIS IS YOUR FINAL WARNING!
Thom Metzger

FRIENDLY FIRE
Bob Black

CALIBAN AND THE WITCHES
Silvia Federici

FIRST AND LAST EMPERORS
The Absolute State & the Body of the Despot
Kenneth Dean & Brian Massumi

SHOWER OF STARS
The Initiatic Dream in Sufism & Taoism
Peter Lamborn Wilson

THIS WORLD WE MUST LEAVE
& OTHER ESSAYS
Jacques Camatte

PIRATE UTOPIAS
Moorish Corsairs & European Renegadoes
Peter Lamborn Wilson

40TH CENTURY MAN
Andy Clausen

FLESH MACHINE
Designer Babies & The Politics of New Genetics
Critical Art Ensemble

WIGGLING WISHBONE
Stories of Patasexual Speculation
Bart Plantenga

FUTURE PRIMITIVE AND OTHER ESSAYS
John Zerzan

THE ELECTRONIC DISTURBANCE
Critical Art Ensemble

X TEXTS
Derek Pell

WHORE CARNIVAL
Shannon Bell, ed.

CRIMES OF CULTURE
Richard Kostelanetz

INVISIBLE GOVERNANCE
The Art of African Micropolitics
David Hecht & Maliqalim Simone

THE LIZARD CLUB
Steve Abbott

CRACKING THE MOVEMENT
Squatting Beyond the Media
Foundation for Advancement of Illegal Knowledge

SOCIAL OVERLOAD
Henri-Pierre Jeudy

ELECTRONIC CIVIL DISOBEDIENCE
Critical Art Ensemble

SEMIOTEXT(E), THE JOURNAL
Jim Fleming & Sylvère Lotringer, Editors

POLYSEXUALITY
François Peraldi, ed.

OASIS
Timothy Maliqalim Simone, et al., eds.

SEMIOTEXT(E) USA
Jim Fleming & Peter Lamborn Wilson, eds.

SEMIOTEXT(E) ARCHITECTURE
Hraztan Zeitlian, ed.

SEMIOTEXT(E) SF
*Rudy Rucker, Robert Anton Wilson,
Peter Lamborn Wilson, eds.*

RADIOTEXT(E)
Neil Strauss & Dave Mandl, eds.

SEMIOTEXT(E) CANADAS
Jordan Zinovich, ed.

PLOVER PRESS

THE COURAGE TO STAND ALONE
U. G. Krishnamurti

THE MOTHER OF GOD
Luna Tarlo

AUTONOMEDIA DISTRIBUTION

DRUNKEN BOAT
An Anarchist Review of Literature & the Arts
Max Blechman, ed.

LUSITANIA
A Journal of Reflection & Oceanography
Martim Avillez, ed.

FELIX
The Review of Television & Video Culture
Kathy High, ed.

RACE TRAITOR
A Journal of the New Abolitionism
John Garvey & Noel Ignatiev, eds.

XXX FRUIT
Anne D'Adesky, ed.

BENEATH THE EMPIRE OF THE BIRDS
Carl Watson

LIVING IN VOLKSWAGEN BUSES
Julian Beck

I SHOT MUSSOLINI
Elden Garnet

ANARCHY AFTER LEFTISM
Bob Black

ALL COTTON BRIEFS
M. Kasper

BELLE CATASTROPHE
Carl Watson & Shalom